GENERALIZED INVERSES OF LINEAR OPERATORS

PURE AND APPLIED MATHEMATICS

A Program of Monographs, Textbooks, and Lecture Notes

EXECUTIVE EDITORS—MONOGRAPHS, TEXTBOOKS, AND LECTURE NOTES

Earl J. Taft
Rutgers University
New Brunswick, New Jersey

Edwin Hewitt
University of Washington
Seattle, Washington

CHAIRMAN OF THE EDITORIAL BOARD

S. Kobayashi
University of California, Berkeley
Berkeley, California

EDITORIAL BOARD

Masanao Aoki
University of California, Los Angeles

W. S. Massey
Yale University

Glen E. Bredon
Rutgers University

Irving Reiner
University of Illinois at Urbana-Champaign

Sigurdur Helgason
Massachusetts Institute of Technology

Paul J. Sally, Jr.
University of Chicago

G. Leitman
University of California, Berkeley

Jane Cronin Scanlon
Rutgers University

Marvin Marcus
University of California, Santa Barbara

Martin Schechter
Yeshiva University

Julius L. Shaneson
Rutgers University

MONOGRAPHS AND TEXTBOOKS IN PURE AND APPLIED MATHEMATICS

1. K. Yano. Integral Formulas in Riemannian Geometry (1970)
2. S. Kobayashi. Hyperbolic Manifolds and Holomorphic Mappings (1970)
3. V. S. Vladimirov. Equations of Mathematical Physics (A. Jeffrey, editor; A. Littlewood, translator) (1970)
4. B. N. Pshenichnyi. Necessary Conditions for an Extremum (L. Neustadt, translation editor; K. Makowski, translator) (1971)
5. L. Narici, E. Beckenstein, and G. Bachman. Functional Analysis and Valuation Theory (1971)
6. D. S. Passman. Infinite Group Rings (1971)
7. L. Dornhoff. Group Representation Theory (in two parts). Part A: Ordinary Representation Theory. Part B: Modular Representation Theory (1971, 1972)
8. W. Boothby and G. L. Weiss (eds.). Symmetric Spaces: Short Courses Presented at Washington University (1972)
9. Y. Matsushima. Differentiable Manifolds (E. T. Kobayashi, translator) (1972)
10. L. E. Ward, Jr. Topology: An Outline for a First Course (1972)
11. A. Babakhanian. Cohomological Methods in Group Theory (1972)
12. R. Gilmer. Multiplicative Ideal Theory (1972)
13. J. Yeh. Stochastic Processes and the Wiener Integral (1973)
14. J. Barros-Neto. Introduction to the Theory of Distributions (1973)
15. R. Larsen. Functional Analysis: An Introduction (1973)
16. K. Yano and S. Ishihara. Tangent and Cotangent Bundles: Differential Geometry (1973)
17. C. Procesi. Rings with Polynomial Identities (1973)
18. R. Hermann. Geometry, Physics, and Systems (1973)
19. N. R. Wallach. Harmonic Analysis on Homogeneous Spaces (1973)
20. J. Dieudonné. Introduction to the Theory of Formal Groups (1973)
21. I. Vaisman. Cohomology and Differential Forms (1973)
22. B.-Y. Chen. Geometry of Submanifolds (1973)
23. M. Marcus. Finite Dimensional Multilinear Algebra (in two parts) (1973, 1975)
24. R. Larsen. Banach Algebras: An Introduction (1973)
25. R. O. Kujala and A. L. Vitter (eds). Value Distribution Theory: Part A; Part B. Deficit and Bezout Estimates by Wilhelm Stoll (1973)
26. K. B. Stolarsky. Algebraic Numbers and Diophantine Approximation (1974)
27. A. R. Magid. The Separable Galois Theory of Commutative Rings (1974)
28. B. R. McDonald. Finite Rings with Identity (1974)
29. I. Satake. Linear Algebra (S. Koh, T. Akiba, and S. Ihara, translators) (1975)
30. J. S. Golan. Localization of Noncommutative Rings (1975)
31. G. Klambauer. Mathematical Analysis (1975)
32. M. K. Agoston. Algebraic Topology: A First Course (1976)
33. K. R. Goodearl. Ring Theory: Nonsingular Rings and Modules (1976)
34. L. E. Mansfield. Linear Algebra with Geometric Applications (1976)
35. N.J. Pullman. Matrix Theory and its Applications: Selected Topics (1976)
36. B. R. McDonald. Geometric Algebra Over Local Rings (1976)
37. C. W. Groetsch. Generalized Inverses of Linear Operators: Representation and Approximation (1977)

GENERALIZED INVERSES OF LINEAR OPERATORS
REPRESENTATION AND APPROXIMATION

C. W. Groetsch
Department of Mathematical Sciences
University of Cincinnati
Cincinnati, Ohio

MARCEL DEKKER, INC.   New York and Basel

Library of Congress Cataloging in Publication Data

Groetsch, C   W
   Generalized inverses of linear operators.

   Bibliography: p.
   Includes index.
   1. Linear operators—Generalized inverses.
2. Approximation theory. 3. Hilbert space. I. Title.
QA329.2.G76   515'.72   77-3864
ISBN 0-8247-6615-6

COPYRIGHT © 1977 by MARCEL DEKKER, INC.   ALL RIGHTS RESERVED.

Neither this book nor any part may be reproduced or transmitted in any form or by any means, electronic or mechanical, including photocopying, microfilming, and recording, or by any information storage and retrieval system, without permission in writing from the publisher.

MARCEL DEKKER, INC.
270 Madison Avenue, New York, New York 10016

Current printing (last digit):
10 9 8 7 6 5 4 3 2 1

PRINTED IN THE UNITED STATES OF AMERICA

> You must always invert.
>
> C. G. J. Jacobi

To Sandra, Kurt and Heidi

PREFACE

This monograph stems from lectures which I gave during the years 1973-1975 at the University of Cincinnati and the University of Rhode Island. The aim of these lectures was to present a unified treatment of the approximation theory of generalized inverses of bounded linear operators in Hilbert space. General representation theorems which unify many diverse computational procedures appear here for the first time in book form and a new unified treatment of error bounds is presented. In addition much of the recent literature on approximation methods for generalized inverses of linear operators is surveyed.

Recently several books on the theory and application of matrix generalized inverses have appeared. Particular mention should be made of the excellent volume by Ben-Israel and Greville which contains a chapter on generalized inverses of linear operators in Hilbert space and the recently published book edited by Nashed. However, at the present time, I believe this to be the only self-contained treatment of the computational theory of generalized inverses of bounded linear operators in Hilbert space which is suitable for use as a text in a first or second year graduate seminar. It is also hoped that this text will be useful as supplemental reading in courses on linear operator theory and advanced numerical analysis. With these uses in mind, a list of exercises is provided at the end of each chapter.

The chapters are divided into sections, and equations to which we will have occasion to refer are numbered consecutively in each section. Every proposition has a unique identifying number; the number "a.b.c." refers to proposition number "c" in section "b" of chapter "a". The symbol "#" is used to indicate the end of a proof.

Many friends, colleagues and students have read and commented on various parts of the manuscript. Thanks are particularly due to Betsy Conway, Mohab El-Samaloty, Bart Jacobs, Alan Lazer, John Montgomery, Lew Pakula and Ghasi Verma. I owe a special debt of gratitude to Professor Marvin Marcus for many valuable suggestions on content and style. Finally, I wish to express my appreciation to Linda Patterson for an efficient job of typing. The usual statement concerning the ultimate responsibility for residual errors applies here.

C. W. Groetsch

CONTENTS

I   HILBERT SPACE AND HILBERT SPACE OPERATORS

      § 1   Hilbert space   1

      § 2   Linear operators   9

      § 3   Spectral theory   17

      § 4   Differentiation   24

             Exercises   32

II   GENERALIZED INVERSES OF BOUNDED LINEAR OPERATORS WITH CLOSED RANGE

      § 1   Definition and basic properties   37

      § 2   Other definitions   47

      § 3   A representation theorem and applications   54

             1.   An operator-valued integral representation   59

             2.   An iterative method   62

             3.   An analog of Schulz's method   69

             4.   Hyperpower methods   71

             5.   Tihonov regularization   74

             6.   A method based on interpolatory function theory   76

             7.   Other representations   81

      § 4   Steepest descent   82

      § 5   The conjugate gradient method   90

             Exercises   110

III   GENERALIZED INVERSES OF BOUNDED LINEAR OPERATORS WITH ARBITRARY RANGE

      § 1   Definition and basic properties   113

	§ 2	A representation theorem	119
	§ 3	Steepest descent	124
	§ 4	The conjugate gradient method	130
		Exercises	143

BIBLIOGRAPHY — 145
INDEX OF SYMBOLS — 162
SUBJECT INDEX — 163

GENERALIZED INVERSES OF LINEAR OPERATORS

CHAPTER I

HILBERT SPACE AND HILBERT SPACE OPERATORS

Hilbert space, since it is the most natural infinite dimensional structure into which our geometric intuition generally carries over, has long been considered as the appropriate vehicle for the study of many linear problems. Although we assume that the reader has some familiarity with Hilbert space and the theory of linear operators, we have included this preliminary chapter in an effort to establish coherent notation and make the presentation reasonably self-contained. It is hoped that this chapter will serve as a brief introduction to basic Hilbert space theory. The results in this chapter are for the most part stated without proof. Proofs of many of the theorems are outlined in the exercises and references are given to standard books on functional analysis and operator theory where more leisurely accounts of the basic theory can be found.

SECTION 1
HILBERT SPACE

We assume that the reader is familiar with the concept of a linear space. By a normed linear space we will mean a linear space E endowed with a real-valued function $||\cdot||$ which satisfies the following axioms

$$||x|| \geq 0; \quad ||x|| = 0 \text{ if and only if } x = 0,$$
$$||\alpha x|| = |\alpha| \, ||x||,$$
$$||x + y|| \leq ||x|| + ||y||,$$

where x and y are arbitrary elements of E and α is any (real or complex, depending on the context) scalar.

Definition. An <u>inner product space</u> (also called a pre-Hilbert space) is a linear space E endowed with a scalar valued function $(\cdot,\cdot)$, called an <u>inner product</u> for E, which satisfies for any $x,y,z \in E$ and any scalars α and β

$$(x,y) = \overline{(y,x)}$$
$$(\alpha x + \beta y, z) = \alpha(x,z) + \beta(y,z)$$
$$(x,x) \geq 0, \text{ with equality only if } x = 0.$$

The bar denotes complex conjugation (of course, if the scalar field is taken to be the real numbers then the first axiom becomes $(x,y) = (y,x)$). By using Theorem 1.1.1, it is easy to show that any inner product space is also a normed linear space where the norm is defined by

$$||x|| = (x,x)^{1/2}.$$

HILBERT SPACE

The next theorem gives an inequality which is very basic in the study of inner product spaces; its proof is outlined in the exercises.

__Theorem 1.1.1.__ (Schwarz's inequality) For any x and y in an inner product space E

$$|(x,y)| \leq ||x||\,||y||$$

with equality holding if and only if x and y are linearly dependent (i.e., there exist scalars α and β, not both equal to zero, such that $\alpha x + \beta y = 0$).

Given points x and y in a normed linear space E the <u>distance</u> between x and y is defined as $d(x,y) = ||x - y||$. It is easy to see that d defines a metric on E which in turn generates a topology for E (the norm topology). Schwarz's inequality shows that in an inner product space the function $(\cdot,\cdot)$ is continuous in the product topology on $E \times E$ induced by the norm topology on E.

The proof of the following theorem is routine.

__Theorem 1.1.2.__ If E is an inner product space then
(a) $||x + y||^2 + ||x - y||^2 = 2(||x||^2 + ||y||^2)$
(b_1) $(x,y) = (||x + y||^2 - ||x - y||^2)/4$ (real scalars)
(b_2) $(x,y) = (||x + y||^2 - ||x - y||^2 +$
         $i||x + iy||^2 - i||x - iy||^2)/4$ (complex scalars)

Identity (a) above is called the __parallelogram law__ owing to its obvious geometrical interpretation. The identities (b) are usually referred to in the literature as the __polarization identities__.

__Definition__. A sequence $\{x_n\}$ in a normed linear space is called a __Cauchy sequence__ if given $\varepsilon > 0$ there is a positive integer $N(\varepsilon)$ such that $n,m > N(\varepsilon)$ implies $||x_n - x_m|| < \varepsilon$.

__Definition__. A normed linear space E is said to be __complete__ (also called a Banach space) if every Cauchy sequence in E converges to some point of E.

__Definition__. A complete inner product space is called a __Hilbert space__.

The most important examples of Hilbert spaces are the real Hilbert spaces

$$\ell^2 = \{\{x_i\} : x_i \in R \text{ and } \sum_{i=1}^{\infty} x_i^2 < \infty\}$$

where the inner product of two vectors $x = \{x_i\}$ and $y = \{y_i\}$ is defined by

$$(x,y) = \sum_{i=1}^{\infty} x_i y_i,$$

HILBERT SPACE

and the Hilbert space $L^2[a,b]$ consisting of all (equivalence classes of) square integrable functions on $[a,b]$ with the inner product

$$(f,g) = \int_a^b f(t)g(t)\, dt.$$

Of course, with minor modifications in the above descriptions we could define complex versions of the spaces ℓ^2 and $L^2[a,b]$. It goes without saying that the finite dimensional spaces R^n and $\mathbb{C}^n$ are Hilbert spaces under the usual inner products.

<u>Definition</u>. A scalar valued linear function defined on a linear space E is called a <u>linear functional</u> on E (i.e., $f : E \to F$, where F is the scalar field, is a linear functional if $f(\alpha x + \beta y) = \alpha f(x) + \beta f(y)$, for $\alpha, \beta \in F$ and $x, y \in E$).

If addition and scalar multiplication are defined on the set of linear functions by

$$(\phi + \psi)(x) = \phi(x) + \psi(x)$$
$$(\alpha \phi)(x) = \alpha \phi(x)$$

we see that this set forms a linear space. The linear space of all (norm) continuous linear functionals on a normed linear space E is denoted by E^* and is called the <u>dual space</u> of E. The space E^* is a normed linear space where the norm is defined for $\phi \in E^*$ by

$$||\phi|| = \sup\{|\phi(x)| : x \in E, ||x|| = 1\}.$$

If the space E is complete then so is E^*. The next theorem shows that Hilbert space enjoys the property of being self-dual (for a proof see the exercises).

<u>Theorem 1.1.3</u>. (Riesz Representation Theorem) If ϕ is a continuous linear functional on a Hilbert space H, then there exists a unique $y \in H$ such that

$$\phi(x) = (x,y)$$

for each $x \in H$.

<u>Definition</u>. A subset S of a linear space E is called a <u>subspace</u> of E if S is itself a linear space. If E is a normed linear space then a subspace S of E is called a <u>closed</u> <u>subspace</u> if S is closed in the norm topology for E.

We note that a closed subspace of a complete linear space is itself complete. We now introduce the concept of convexity, which plays a fundamental role in linear space theory.

<u>Definition</u>. A subset C of a normed linear space is called convex if $tx + (1 - t)y \in C$ for all $x,y \in C$ and all $t \in [0,1]$.

HILBERT SPACE

The next theorem is a basic principal in the theory of optimization and best approximation.

Theorem 1.1.4. A closed convex subset C of a Hilbert space contains a unique vector of smallest norm.

Proof. Let $M = \inf\{||x|| : x \in C\}$ and choose a sequence $\{x_n\} \subset C$ such that $\lim_n ||x_n|| = M$. Since C is convex, we have by use of the parallelogram law

$$||x_n - x_m||^2 = 2(||x_n||^2 + ||x_m||^2) - 4||(x_n + x_m)/2||^2$$
$$\leq 2(||x_n||^2 + ||x_m||^2) - 4M^2$$

which converges to 0 as $n,m \to \infty$. Hence $\{x_n\}$ is a Cauchy sequence which therefore has a limit $x \in C$ (since C is closed). It is now easy to see that x is the unique vector in C of minimal norm. Indeed, if y is a vector in C with norm M which is distinct from x, then

$$0 < ||x - y||^2 = 4M - 4||(x + y)/2||^2.$$

Therefore,

$$||(x + y)/2|| < M$$

which is a contradiction. #

The following is an easy consequence of (1.1.4).

Corollary 1.1.5. If C is a closed convex subset of a Hilbert space H, then for each $u \in H$ there is a unique $x \in C$ such that

$$||u - x|| = \inf\{||u - y|| : y \in C\}.$$

Inner product spaces have a richer geometrical structure than general linear spaces owing to the fact that in spaces with an inner product the concept of perpendicularity can be developed.

Definition. Two vectors x and y in an inner product space are said to be <u>orthogonal</u>, denoted $x \perp y$, if $(x,y) = 0$. If S is a subspace of an inner product space E we define its <u>orthogonal complement</u> by

$$S^{\perp} = \{y \in E : x \perp y \text{ for all } x \in S\}.$$

Note that if x and y are orthogonal then they satisfy the <u>Pythagorean property</u>

$$||x + y||^2 = ||x||^2 + ||y||^2.$$

The name is suggestive of the geometrical significance of this identity. It is worthwhile to point out that since the inner product is continuous it follows that for any subspace S the

HILBERT SPACE

set $S^\perp$ is a <u>closed</u> subspace. It is of fundamental importance that in a Hilbert space a closed subspace and its orthogonal complement decompose the space in the following sense.

<u>Theorem 1.1.6</u>. If S is a closed subspace of a Hilbert space H, then H can be written as the direct sum of S and $S^\perp$, denoted $H = S \oplus S^\perp$, meaning that each $x \in H$ can be written uniquely as $x = x_1 + x_2$, where $x_1 \in S$ and $x_2 \in S^\perp$.

<u>Proof</u>. Given $x \in H$ there exists by (1.1.5) a unique vector $x_1 \in S$ such that $||x - x_1||$ is minimal. Let $x_2 = x - x_1$. For any $y \in S$ with $||y|| = 1$ we have

$$||x_2||^2 = ||x - x_1||^2 \leq ||x - x_1 - (x_2,y)y||^2$$
$$= ||x_2||^2 - |(x_2,y)|^2 \leq ||x_2||^2.$$

Therefore, $(x_2,y) = 0$, and it follows that $x_2 \in S^\perp$. Hence we see that $x = x_1 + x_2 \in S \oplus S^\perp$. Uniqueness of the representation is easy to establish. #

SECTION 2
<u>LINEAR OPERATORS</u>

This section sets forth some of the salient features of the theory of linear operators. The results given here will form a foundation for our general development of generalized inverses of bounded linear operators.

<u>Definition</u>. A function T defined on a linear space E_1 and taking values in a linear space E_2 is called a <u>linear operator</u> if

$$T(\alpha x + \beta y) = \alpha Tx + \beta Ty$$

for any $x, y \in E_1$ and any scalars α and β. When E_1 and E_2 are normed linear spaces with norms $||\cdot||_1$ and $||\cdot||_2$ respectively, we say that a linear operator T from E_1 to E_2 is <u>bounded</u> if the number

$$||T|| = \sup\{||Tx||_2 : ||x||_1 = 1\}$$

is finite.

The number $||T||$ defined above is called the <u>norm</u> of T and the set $L(E_1, E_2)$ of all bounded linear operators from E_1 to E_2, with addition of operators and multiplication of an operator by a scalar defined by

$$(T + S)x = Tx + Sx \quad \text{and} \quad (\alpha T)x = \alpha Tx$$

is itself a normed linear space under this norm. We recall that bounded linear operators are precisely the linear operators which are continuous from the norm topology on E_1 to the norm topology on E_2.

HILBERT SPACE

We now introduce two natural notions of convergence for sequences of linear operators.

<u>Definition</u>. Suppose E_1 and E_2 are normed linear spaces and $\{T_n\}$ is a sequence of linear operators in $L(E_1,E_2)$. The sequence $\{T_n\}$ is said to <u>converge uniformly</u> to $T \in L(E_1,E_2)$ if

$$\lim_n ||T_n - T|| = 0.$$

The sequence is said to <u>converge strongly</u> to $T \in L(E_1,E_2)$ if

$$\lim_n ||T_n x - Tx||_2 = 0$$

for each $x \in E_1$.

Note that for each $x \in H_1$,

$$||Tx||_2 \leq ||T||\, ||x||_1$$

and hence if $\{T_n\}$ converges uniformly to T, then $\{T_n\}$ converges strongly to T. However the converse does not hold.

<u>Definition</u>. If $T \in L(E_1,E_2)$, the <u>range</u> and <u>nullspace</u> of T are defined by

$$R(T) = \{y \in E_2 : y = Tx \quad \text{for some } x \in E_1\}$$

and

$$N(T) = \{x \in E_1 : Tx = 0\}, \quad \text{respectively.}$$

The range and nullspace of $T \in L(E_1,E_2)$ are linear subspaces of E_2 and E_1 respectively. The nullspace is always closed in the topology of E_1.

If H_1 and H_2 are Hilbert spaces with inner products $(\cdot,\cdot)_1$ and $(\cdot,\cdot)_2$ respectively and $T \in L(H_1,H_2)$, then given $y \in H_2$ the linear functional defined on H_1 by

$$\phi(x) = (Tx,y)_2$$

is continuous and hence by the Riesz Theorem there is a vector $z_y \in H_1$ such that

$$\phi(x) = (x,z_y)_1.$$

The mapping $y \to z_y$ is clearly linear. It is also continuous, for setting $x = z_{y_1} - z_{y_2}$ in the identity

$$(x, z_{y_1} - z_{y_2})_1 = (Tx, y_1 - y_2)_2$$

and using Schwarz's inequality we obtain

$$\|z_{y_1} - z_{y_2}\|_1 \leq \|T\|\,\|y_1 - y_2\|.$$

This mapping is so crucial in the theory of linear operators that it warrants special terminology and notation.

HILBERT SPACE

Definition. If H_1 and H_2 are Hilbert spaces with inner products $(\cdot,\cdot)_1$ and $(\cdot,\cdot)_2$ respectively and $T \in L(H_1,H_2)$, then the <u>adjoint</u> of T, denoted T^*, is the unique linear operator in $L(H_2,H_1)$ satisfying

$$(Tx,y)_2 = (x,T^*y)_1$$

for all $x \in H_1$ and $y \in H_2$.

The subscripts on inner product and norm symbols will henceforth be omitted in adherence to a long established tradition of abuse of notation and we trust that no confusion will result. We note the following simple properties of the adjoint operation:

$$(ST)^* = T^*S^*$$
$$(T^*)^* = T$$
$$(T + S)^* = T^* + S^*$$
$$(\alpha T)^* = \overline{\alpha}T^*$$
$$||T^*|| = ||T||$$
$$||T^*T|| = ||TT^*|| = ||T||^2.$$

An operator $T \in L(H,H)$ is called <u>self-adjoint</u> if $T = T^*$. The next theorem establishes fundamental relationships between the range, nullspace and adjoint which will be very useful in the sequel. A proof is outlined in the exercises.

<u>Theorem 1.2.1</u>. If $T \in L(H_1,H_2)$, where H_1 and H_2 are Hilbert

spaces, then

$$N(T)^\perp = \overline{R(T^*)}, \quad R(T)^\perp = N(T^*),$$
$$N(T^*)^\perp = \overline{R(T)}, \quad R(T^*)^\perp = N(T),$$

where the bars denote the (norm) closure of the given subspaces.

If $T \in L(E_1, E_2)$ is a one-to-one operator ($N(T) = \{0\}$), then the inverse mapping T^{-1} defined from $R(T)$ to E_1 is also linear. The next theorem is Banach's celebrated theorem on the inverse mapping. For a proof see Lorch [1, page 38].

<u>Theorem 1.2.2</u>. If E_1 and E_2 are complete normed linear spaces and $T \in L(E_1, E_2)$ is one-to-one and onto then T^{-1} is continuous.

Henceforth we will say that T is <u>invertible</u> if T^{-1} exists and is continuous. If T is a linear operator defined on a linear space E and S is a subspace of E, then $T|S$ will denote the <u>restriction</u> of T to S.

<u>Theorem 1.2.3</u>. Let E_1 and E_2 be complete linear spaces and suppose $T \in L(E_1, E_2)$ has closed range. Then there is a number $m > 0$ such that $||Tx|| \geq m||x||$ for all $x \in N(T)^\perp$.

<u>Proof</u>. Note that $N(T)^\perp$ is complete and $T|N(T)^\perp$ is a one-to-one operator mapping $N(T)^\perp$ onto the complete space $R(T)$. There-

HILBERT SPACE   15

fore by (1.2.2) $[T|N(T)^\perp]^{-1} \in L(R(T),N(T)^\perp)$ and we need only set $m = ||[T|N(T)^\perp]^{-1}||^{-1}$. #

Suppose that S is a closed subspace of a Hilbert space H. Then by (1.1.6) each $x \in H$ can be written uniquely as $x = x_1 + x_2 \in S \oplus S^\perp$. The operator $P \in L(H,S)$ defined by $Px = x_1$ is called the <u>projection</u> of H onto S. Note that P satisfies

$$P = P^*, \quad P = P^2 \text{ and } ||P|| = 1.$$

If P projects H onto S then I - P is the projection of H onto $S^\perp$, where I is the identity operator on H. When we wish to emphasize the relation of P to S we write $P = P_S$. Any operator $T \in L(H,H)$ which satisfies $T = T^2$ and $T = T^*$ is the projection operator of H onto the closed subspace defined by $\{x \in H : Tx = x\}$. Note that the projection operator onto the closed subspace S satisfies the variational property

$$||x - P_S x|| = \inf\{||x - y|| : y \in S\}.$$

<u>Theorem 1.2.4</u>. If H_1 and H_2 are Hilbert spaces and $T \in L(H_1,H_2)$, then R(T) is closed if and only if $R(T^*)$ is closed.

<u>Proof</u>. If R(T) is closed, then by (1.2.3) there is an $m > 0$ such that

16   GENERALIZED INVERSES

$$||Tx|| \geq m||x||, \quad \text{for } x \in N(T)^\perp.$$

Hence if P is the projection of H_1 onto $N(T)^\perp$, then

$$||Tx|| \geq m||Px||, \quad \text{for } x \in H_1.$$

If $y \in \overline{R(T^*)} = N(T)^\perp$ by (1.2.1), then the linear functional ϕ defined on R(T) by $\phi(Tx) = (x,y)$ is bounded:

$$\begin{aligned}|\phi(Tx)| &= |(x,y)| \\ &\leq ||Px||\ ||y|| \\ &\leq m^{-1}||Tx||\ ||y||.\end{aligned}$$

By the Riesz Theorem (1.1.3) there is a vector $Tz \in R(T)$ such that

$$(x,y) = \phi(Tx) = (Tx,Tz)$$

for all $x \in H_1$, but this implies

$$y = T^*Tz \in R(T^*).$$

The converse follows since $T^{**} = T$. #

HILBERT SPACE

SECTION 3
SPECTRAL THEORY

In this section we give a very brief sketch of the spectral theory of bounded self-adjoint operators. The treatment given here is descriptive and heuristic. More detailed and rigorous accounts of the theory can be found in the books of Riesz and Sz. Nagy [1], Taylor [1] and Liusternik and Sobolev [1] and the very readable exposition of Lorch [2].

Definition. Suppose $T \in L(H,H)$ where H is a Hilbert space. A complex number λ is called an **eigenvalue** of T if there is a nonzero vector $x \in H$ such that

$$Tx = \lambda x.$$

Any such vector x is called an **eigenvector** of T associated with the eigenvalue λ.

If λ is not an eigenvalue of T then the inverse mapping $(T - \lambda I)^{-1}$ can be defined on $R(T - \lambda I)$. This mapping is linear but it may not be bounded. The set of λ for which $(T - \lambda I)^{-1}$ is defined on all of H and is bounded has special significance.

Definition. If $T \in L(H,H)$ where H is a Hilbert space, then the **resolvent set** of T, denoted $\rho(T)$, is defined by

$$\rho(T) = \{\lambda \in \mathbb{C} : (T - \lambda I)^{-1} \in L(H,H)\}.$$

The <u>spectrum</u> of T, denoted $\sigma(T)$, is the complement of $\rho(T)$,

$$\sigma(T) = \mathbb{C} - \rho(T).$$

The <u>spectral radius</u> of T, denoted $|\sigma(T)|$, is defined by

$$|\sigma(T)| = \sup\{|\lambda| : \lambda \in \sigma(T)\}.$$

Note that each eigenvalue of T is a point of $\sigma(T)$. The spectrum of an operator $T \in L(H,H)$ is compact and $|\sigma(T)| \leq ||T||$ (see exercise 10).

<u>Definition</u>. A self-adjoint operator $T \in L(H,H)$ is called <u>nonnegative</u>, denoted $T \geq 0$, if

$$(Tx,x) \geq 0$$

for each $x \in H$. If the strict inequality holds for each nonzero x, we say that T is <u>positive</u>, denoted $T > 0$.

If A and T are self-adjoint operators we write $A \geq T$ if $A - T \geq 0$ and $A > T$ if $A - T > 0$.

<u>Theorem 1.3.1</u>. The spectrum of a self-adjoint operator T is real; furthermore, if a and b are real numbers and $aI \leq T \leq bI$, then $\sigma(T) \subset [a,b]$.

HILBERT SPACE

Proof. Suppose $\lambda = \alpha + \beta i$, where $\beta \neq 0$. We then have for any $x \in H$,

$$||Tx - \lambda x||^2 = (Tx - \alpha x - \beta i x, Tx - \alpha x - \beta i x)$$
$$= ||Tx - \alpha x||^2 + \beta^2 ||x||^2, \text{ since } T = T^*$$
$$\geq \beta^2 ||x||^2.$$

Therefore $N(T - \lambda I) = \{0\}$. Similarly $N(T - \bar{\lambda} I) = \{0\}$ and hence by (1.2.1)

$$\overline{R(T - \lambda I)} = N(T - \lambda I)^{\perp} = \{0\}^{\perp} = H.$$

We therefore see that $T - \lambda I$ is a one-to-one operator with a dense range. Suppose that $y \in H$, then there is a sequence

$$Tx_n - \lambda x_n \in R(T - \lambda I)$$

such that $Tx_n - \lambda x_n \to y$ as $n \to \infty$. By setting $x = x_n - x_m$ in the inequality above we find that

$$||(T - \lambda I)(x_n - x_m)||^2 \geq \beta^2 ||x_n - x_m||^2.$$

Since $\{(T - \lambda I)x_n\}$ is convergent it follows that $\{x_n\}$ is a Cauchy sequence and hence has a limit $x \in H$. By continuity we see that $Tx - \lambda x = y$ and therefore $R(T - \lambda I) = H$. Since $T - \lambda I$ is one-to-one and onto it is invertible by (1.2.2) and hence $\lambda \in \rho(T)$. We have therefore shown that if $\lambda \in \sigma(T)$ then λ must be real.

If $aI \leq T$ and $\lambda < a$ then we have

$$||Tx - \lambda x|| \, ||x|| \geq ((T - \lambda I)x,x) \geq (a - \lambda)||x||^2.$$

The same argument given above now shows that $\lambda \in \rho(T)$. Similarly if $T \leq bI$ and $b < \lambda$ then $\lambda \in \rho(T)$. Consequently, if $aI \leq T \leq bI$, then $\sigma(T) \subset [a,b]$. #

Suppose that T is a real symmetric $n \times n$ matrix (considered as an operator on R^n) and that $\lambda_1, \ldots, \lambda_n$ are the eigenvalues of T (we assume for simplicity of the exposition that the eigenvalues are distinct). It is easy to show that eigenvectors associated with distinct eigenvalues are orthogonal. Indeed if u_1 and u_2 are eigenvectors associated with eigenvalues λ_1 and λ_2 respectively, then since T is self-adjoint, λ_1 and λ_2 are real and

$$\lambda_1(u_1,u_2) = (Tu_1,u_2) = (u_1,Tu_2) = \lambda_2(u_1,u_2)$$

from which it follows that u_1 and u_2 must be orthogonal. Let $u_1, \ldots, u_n$ be eigenvectors of norm 1 associated with the eigenvalues $\lambda_1, \ldots, \lambda_n$. Then $\{u_1, \ldots, u_n\}$ forms an orthonormal basis for R^n and each $x \in R^n$ may be written

HILBERT SPACE

$$x = (x,u_1)u_1 + \cdots + (x,u_n)u_n.$$

We assume that the eigenvalues are ordered in the following way

$$\lambda_1 < \lambda_2 < \cdots < \lambda_n.$$

We define the <u>resolution of the identity</u> (or spectral family) corresponding to T as the projection valued function E_λ of the real variable λ which satisfies

$$E_\lambda x = 0 \quad \text{for } \lambda \leq \lambda_1,$$
$$E_\lambda = P_{S_i} \quad \text{for } \lambda_i < \lambda \leq \lambda_{i+1},$$
$$E_\lambda = I \quad \text{for } \lambda_n < \lambda,$$

where S_i is the subspace spanned by $\{u_1,\ldots,u_i\}$. Notice that the function E_λ is left continuous, i.e., $\lim_{\lambda \to \mu^-} E_\lambda = E_\mu$. If we define the projection $\Delta_i E$ by $\Delta_1 E = P_{S_1}$ and

$$\Delta_i E x = (E_{\lambda_i} - E_{\lambda_{i-1}})x = (x,u_i)u_i, \quad i = 2,\ldots,n,$$

then we see that

$$Tx = \lambda_1(x,u_1)u_1 + \lambda_2(x,u_2)u_2 + \cdots + \lambda_n(x,u_n)u_n$$
$$= \sum_{i=1}^{n} \lambda_i \Delta_i E x.$$

This is essentially the content of the spectral theorem for finite dimensional self-adjoint operators.

<u>Definition</u>. Let H be a Hilbert space. A family of projection operators $\{E_\lambda\}$ indexed by the real parameter λ is called a <u>resolution of the identity</u> for H if

(i) $\lambda < \mu$ implies $E_\lambda \leq E_\mu$,
(ii) $\lambda \to -\infty$ implies $E_\lambda \to 0$ strongly,
(iii) $\lambda \to \infty$ implies $E_\lambda \to I$ strongly,
(iv) $\lim_{\lambda \to \mu^-} E_\lambda = E_\mu$ strongly.

For a proof of the next theorem the reader may consult Lorch [1], Liusternik and Sobolev [1] or other standard references on functional analysis.

<u>Theorem 1.3.2</u>. Every self-adjoint operator $T \in L(H,H)$ generates a resolution of the identity $\{E_\lambda\}$ for H such that

(i)   if $C \in L(H,H)$ and $CT = TC$, then $CE_\lambda = E_\lambda C$ for each real λ;

(ii)   if $aI \leq T \leq bI$, then $E_\lambda = 0$ for $\lambda \leq a$ and $E_\lambda = I$ for $\lambda > b$;

(iii) $T = \int_{\sigma(T)} \lambda \, dE_\lambda$

where the integral (interpreted as the limit of sums of the form

HILBERT SPACE

$\sum \lambda_i \Delta_i E$, where $\Delta_i E = E_{\lambda_i} - E_{\lambda_{i-1}}$ and $\{\lambda_i\}$ is a partition of an interval containing $\sigma(T)$) converges in the uniform operator topology.

The next result is known as the Spectral Mapping Theorem (a proof is outlined in the exercises).

Theorem 1.3.3. If $T \in L(H,H)$ and p is a polynomial, then

$$\sigma(p(T)) = p(\sigma(T)) = \{p(\lambda) : \lambda \in \sigma(T)\}.$$

Note that if p is any polynomial we have by way of the finite dimensional spectral theorem for any symmetric $n \times n$ matrix T,

$$p(T)x = p(\lambda_1)(x,u_1)u_1 + \cdots + p(\lambda_n)(x,u_n)u_n,$$

that is

$$p(T)x = \sum_{i=1}^{n} p(\lambda_i) \Delta_i E.$$

Motivated by this example we see that if $T \in L(H,H)$ is self-adjoint and f is a real-valued continuous function defined on $\sigma(T)$, the spectral theorem allows us to define the operator $f(T) \in L(H,H)$ by

$$f(T) = \int_{\sigma(T)} f(\lambda)\, dE_\lambda$$

where the integral converges in the uniform operator topology. In particular, if T is invertible ($0 \notin \sigma(T)$) then it can be shown that the inverse of T is given by

$$T^{-1} = \int_{\sigma(T)} \lambda^{-1}\, dE_\lambda.$$

Finally, we mention that since $\sigma(T)$ is compact and $\{E_\lambda\}$ is uniformly bounded, if $\{f_\beta(\lambda)\}_{\beta > 0}$ is a family of continuous real-valued functions on $\sigma(T)$ such that

$$\lim_\beta f_\beta(\lambda) = f(\lambda)$$

uniformly on $\sigma(T)$, then

$$\lim_\beta f_\beta(T) = f(T)$$

in the uniform operator topology for $L(H,H)$.

SECTION 4

DIFFERENTIATION

In this section we will briefly discuss some aspects of differentiation of mappings in linear spaces. The basic idea

HILBERT SPACE 25

of the differential calculus of mappings in linear spaces is the local approximation of nonlinear mappings by <u>linear</u> operators (Fréchet derivatives). This allows one to study the local behavior of a nonlinear mapping by using the theory of linear operators.

<u>Definition</u>. Let E_1 and E_2 be normed linear spaces and suppose that X is an open subset of E_1. A function $f : X \to E_2$ is said to be <u>Fréchet differentiable</u> at $x_0 \in X$ if there is an operator $f'(x_0) \in L(E_1, E_2)$ such that

$$\lim_{||h|| \to 0} \frac{||f(x_0 + h) - f(x_0) - f'(x_0)h||}{||h||} = 0.$$

It is easy to show that this definition is unambiguous, that is, there can be at most one operator $f'(x_0) \in L(E_1, E_2)$ satisfying the above condition.

<u>Theorem 1.4.1</u>. If $f : X \to E_2$ is Fréchet differentiable at $x_0 \in X$, then the operator $f'(x_0) \in L(E_1, E_2)$ satisfying the definition above is unique.

<u>Proof</u>. Suppose $T \in L(E_1, E_2)$ also satisfies the definition of the Fréchet derivative of f at x_0, then for any nonzero $x \in E_1$

$$\frac{||Tx - f'(x_0)x||}{||x||} = \lim_{t \to 0} \frac{||Ttx - f'(x_0)tx||}{||tx||}$$

$$= \lim_{t \to 0} \frac{||Ttx - f(x_0+tx) + f(x_0) + f(x_0+tx) - f(x_0) - f'(x_0)tx||}{||tx||}$$

$$\leq \lim_{t \to 0} \frac{||f(x_0+tx)-f(x_0)-Ttx||}{||tx||} + \lim_{t \to 0} \frac{||f(x_0+tx)-f(x_0)-f'(x_0)tx||}{||tx||}$$

$$= 0.$$

Therefore $||Tx - f'(x_0)x|| = 0$ for each $x \in E_1$, i.e., $T = f'(x_0)$. #

We note that if f is differentiable at x_0, then f is continuous at x_0, since

$$||f(x_0+h)-f(x_0)|| \leq ||f(x_0+h)-f(x_0)-f'(x_0)h|| +$$

$$||f'(x_0)|| \, ||h|| \to 0 \text{ as } ||h|| \to 0.$$

It also seems appropriate to point out that for a real function $f : R \to R$ the above definition coincides with the traditional notion of the derivative of a function if we regard $f'(x_0)$ not as a number but rather as the corresponding operator in $L(R,R)$ given by $x \to f'(x_0)x$.

If $f : E_1 \to E_2$ is itself a bounded linear operator then it follows immediately from the definition that $f'(x_0) = f$ for all

HILBERT SPACE

$x_0 \in E_1$. The reader is invited to verify that the operation of Fréchet differentiation is linear, i.e. if $f,g : E_1 \to E_2$ are differentiable at $x_0 \in E_1$, then

$$(f + g)'(x_0) = f'(x_0) + g'(x_0)$$

and

$$(\alpha f)'(x_0) = \alpha\, f'(x_0).$$

The "chain rule" for derivatives also extends to the more general setting.

<u>Theorem 1.4.2.</u> Suppose $g : E_1 \to E_2$ is Fréchet differentiable at $x_0 \in E_1$ and $f : E_2 \to E_3$ is Fréchet differentiable at $g(x_0)$, then $F : E_1 \to E_3$ defined by $F(x) = f(g(x))$ is Fréchet differentiable at x_0 and

$$F'(x_0) = f'(g(x_0))g'(x_0).$$

<u>Proof.</u> If we set

$$\gamma(h) = g(x_0 + h) - g(x_0) - g'(x_0)h$$

and

$$\phi(h) = f(g(x_0) + h) - f(g(x_0)) - f'(g(x_0))h$$

Then $\gamma(0) = \phi(0) = 0$ and

$$\lim_{||h|| \to 0} \frac{||\gamma(h)||}{||h||} = \lim_{||h|| \to 0} \frac{||\phi(h)||}{||h||} = 0.$$

Now

$$F(x_0 + h) - F(x_0) - f'(g(x_0))g'(x_0)h$$
$$= f(g(x_0) + g'(x_0)h + \gamma(h)) - f(g(x_0)) - f'(g(x_0))g'(x_0)h$$
$$= f'(g(x_0))\gamma(h) + \phi(g'(x_0)h + \gamma(h))$$

and

$$\frac{||f'(g(x_0))\gamma(h)||}{||h||} \leq ||f'(g(x_0))|| \frac{||\gamma(h)||}{||h||} \to 0$$

as $||h|| \to 0$. Also, from the definition of ϕ we see that for a given $\varepsilon > 0$ there is a $\delta_1 > 0$ such that $||g'(x_0)h + \gamma(h)|| < \delta_1$ implies

$$||\phi(g'(x_0)h + \gamma(h))|| < \varepsilon ||g'(x_0)h + \gamma(h)||.$$

But since $g'(x_0)$ and $\gamma(h)$ are continuous at 0 there is a $\delta_2 > 0$ such that $||h|| < \delta_2$ implies

$$||g'(x_0)h + \gamma(h)|| < \delta_1.$$

Hence we see that for $||h|| < \delta_2$,

$$||\phi(g'(x_0)h + \gamma(h))|| < \varepsilon(||g'(x_0)|| \, ||h|| + ||\gamma(h)||)$$

HILBERT SPACE 29

and therefore

$$\lim_{||h||\to 0} \frac{||\phi(g'(x_0)h + \gamma(h))||}{||h||} = 0.$$

It now follows that

$$\lim_{||h||\to 0} \frac{||F(x_0 + h) - F(x_0) - f'(g(x_0))g'(x_0)h||}{||h||} = 0,$$

i.e. $F'(x_0) = f'(g(x_0))g'(x_0)$. #

If H is a Hilbert space and $f : H \to R$ is a real-valued Fréchet differentiable function on H, then for any $x_0 \in H$, $f'(x_0)$ is a continuous linear functional on H and hence by the Riesz Theorem (1.1.3) there is a unique vector in H which we will denote by $\nabla f(x_0)$, called the <u>gradient</u> of f at x_0, such that

(1) $\qquad f'(x_0)h = (h, \nabla f(x_0))$

for each $h \in H$.

There is an interesting and important relationship between the gradient and the rate of change of a functional at a given point in a specific direction.

Definition. The <u>directional derivative</u> of a functional $f: H \to R$ at x_0 in the direction $h \neq 0$ is the number

$$(2) \qquad Df(x_0, h) = \lim_{t \to 0^+} \frac{f(x_0 + th) - f(x_0)}{t}.$$

providing this limit exists.

Note that if $f: H \to R$ is Fréchet differentiable at x_0, then f has a directional derivative at x_0 in every direction $h \neq 0$. In fact, if f is Fréchet differentiable at x_0, then for any $h \neq 0$,

$$f(x_0 + th) - f(x_0) = f'(x_0)th + r(x_0, th)$$

where

$$\lim_{t \to 0} \frac{||r(x_0, th)||}{||th||} = 0.$$

Therefore,

$$\lim_{t \to 0^+} \frac{f(x_0 + th) - f(x_0)}{t} = f'(x_0)h + \lim_{t \to 0^+} \frac{r(x_0, th)}{t}$$

$$= f'(x_0)h.$$

It now follows by (1) and (2) that

HILBERT SPACE

(3) $$Df(x_0,h) = (h,\nabla f(x_0)).$$

The reader who is interested in a more complete account with an extensive bibliography of differentiation in abstract spaces is invited to enjoy the survey article of Nashed [3].

EXERCISES

1. Prove Schwarz's inequality. (Hint: First suppose that (x,y) is a real number and that x and y are linearly independent. Then for each real scalar λ,

   $$0 < ||\lambda y - x||^2 = ||y||^2 \lambda^2 - 2(x,y)\lambda + ||x||^2.$$

   It follows that the discriminant of this quadratic form must be positive which gives the result. If $(x,y) = e^{i\theta}|(x,y)|$ for some real θ, then $|(x,y)| = (x',y)$ where $x' = e^{-i\theta}x$ and the previous argument can be used on the real quantity (x',y).)

2. Show that the vector x in (1.1.5) may be characterized as the unique vector in C satisfying

   $\text{Re}(u - x, x - y) \geq 0$, for all $y \in C$.

   Give a geometric interpretation of this condition.

3. A normed linear space $(E, ||\cdot||)$ is called <u>uniformly convex</u> (see Clarkson [1]) if given $\varepsilon > 0$ there is a $\delta = \delta(\varepsilon)$ with $0 < \delta < 1$, such that $||x|| \leq 1$, $||y|| \leq 1$ and $||x - y|| \geq \varepsilon$ imply that

   $||(x + y)/2|| \leq 1 - \delta.$

   Show that every inner product space is uniformly convex.

HILBERT SPACE

4. A normed linear space $(E, ||\cdot||)$ is called <u>strictly convex</u> if $||x|| \leq 1$, $||y|| \leq 1$ and $x \neq y$ imply that

$$||(x + y)/2|| < 1.$$

   Of course, every uniformly convex space is strictly convex.

   (a) Give an example of a finite dimensional space which is not strictly convex.

   (b) Show that every finite dimensional strictly convex space is also uniformly convex. (Hint: Use the fact that the unit ball in a finite dimensional space is compact.)

5. Show that (1.1.4) and (1.1.5) remain valid in a complete uniformly convex space.

6. Show that the projection operator of a Hilbert space onto a closed subspace is self-adjoint.

7. Show that every operator $T \in L(H,H)$ satisfying $T = T^2 = T^*$ is a projection operator.

8. Prove the Riesz Representation Theorem. (Hint: If $\phi(x) = 0$ for all x, take $y = 0$. Otherwise set $N = \{x \in H: \phi(x) = 0\}$ and choose $z \in N^\perp$ such that $\phi(z) = 1$, then $y = z/||z||^2$.)

9. Show that if $S \in L(H_1, H_2)$ and $||I - S|| < 1$, then S is invertible

(in fact, $S^{-1} = \sum_{k=0}^{\infty} (I - S)^k$). Conclude that if A is invertible and $||A - B|| < ||A^{-1}||^{-1}$, then B is invertible.

10. Suppose $T \in L(H,H)$ and $|\lambda| > ||T||$. Use the previous exercise to conclude that $\lambda \in \rho(T)$ and hence $|\sigma(T)| \leq ||T||$. It also follows from the previous exercise that $\rho(T)$ is an open set and hence $\sigma(T)$ is compact.

11. Suppose $A \in L(H,H)$ is invertible and there exist $B, C \in L(H,H)$ such that $AC = BA = I$. Show that $B = C = A^{-1}$.

12. Prove the Spectral Mapping Theorem: $\sigma(p(A)) = p(\sigma(A))$, for each polynomial $p(\lambda)$. (Hint: If λ_o is a fixed complex number then there is a polynomial $q(\lambda)$ such that $p(\lambda) - p(\lambda_o) = (\lambda - \lambda_o)q(\lambda)$. If $(A - \lambda_o I)q(A)$ is invertible, then there is an operator B satisfying $(A - \lambda_o I)q(A)B = B(A - \lambda_o I)q(A) = I$. Verify that $q(A)B = Bq(A) = (A - \lambda_o I)^{-1}$ and conclude that $p(\lambda_o) \in \sigma(p(A))$ if $\lambda_o \in \sigma(A)$. Conversely, for $\mu \in \sigma(p(A))$ we have the factorization $p(\lambda) - \mu = \alpha \prod_{i=1}^{n} (\lambda - \lambda_i)$. Since $p(A) - \mu I$ is not invertible we have $\lambda_k \in \sigma(A)$ for some k and $\mu = p(\lambda_k)$.)

13. Prove (1.2.1). (Hint: $N(T^*) = R(T)^\perp$ is easy. From this we have $R(T^*)^\perp = N(T^{**}) = N(T)$ and $\overline{R(T^*)} = N(T)^\perp$ follows since $S^{\perp\perp} = \overline{S}$ for any set S. The last two relations follow by replacing T by T^* in the previous two.)

HILBERT SPACE

14. Let H_1 and H_2 be real Hilbert spaces and suppose $T \in L(H_1, H_2)$. Define $f: H_1 \to R$ by $f(x) = ||Tx - b||^2$, where $b \in H_2$. Show that $\nabla f(x) = 2(T^*Tx - T^*b)$. In particular, if $f(x) = ||x||^2$ then $\nabla f(x) = 2x$.

15. Suppose that $f: R^3 \to R$ is differentiable at (x_0, y_0, z_0). Show that $\nabla f(x_0, y_0, z_0)$ is given by

    $(\frac{\partial f}{\partial x}(x_0, y_0, z_0), \frac{\partial f}{\partial y}(x_0, y_0, z_0), \frac{\partial f}{\partial z}(x_0, y_0, z_0))$.

16. The result of exercise 9 shows that the set U of invertible operators in $L(H_1, H_2)$ is open in the uniform topology. Suppose $f: U \to L(H_2, H_1)$ is defined by $f(T) = T^{-1}$. Show that

    $f'(T_0)T = T_0^{-1} T T_0^{-1}$.

17. Prove that if E_1 is a normed linear space and E_2 is a Banach space then $L(E_1, E_2)$ is a Banach space.

18. Give an example of a Hilbert space H and a sequence $\{T_n\} \subset L(H, H)$ such that $\{T_n\}$ converges strongly to the zero operator and $||T_n|| = 1$ for all n.

CHAPTER II

GENERALIZED INVERSES OF BOUNDED LINEAR OPERATORS WITH CLOSED RANGE

This chapter is concerned with the definition, representation and approximation of the generalized inverse of an operator $T \in L(H_1, H_2)$, where H_1 and H_2 are Hilbert spaces over the same field of scalars and $R(T)$ is closed. Generalized inverses can be defined in a more abstract setting (see for example Sheffield [1], Nashed and Votruba [1], [2], Caradus [1] and Koliha [2] and the references cited in these works) but it is the opinion of the author that the most natural way to introduce the concept of the generalized inverse is via the variational definition given in this chapter for Hilbert space operators with closed range. Our primary concern, aside from the definition and basic properties, is with general methods for representing and computing generalized inverses. In the first two sections of this chapter we present the definition of the generalized inverse of a bounded linear operator with closed range and show the equivalence of this definition with several other definitions of the generalized inverse. In section 3 we present a general representation theorem (see Groetsch [2]) which encompasses a broad spectrum of specific representations and computational procedures as corollaries. The steepest descent method for pointwise compu-

CLOSED RANGE CASE   37

tation of the generalized inverse of an operator is
taken up in section 4 and in the final section results of
Kammerer and Nashed [2] on the conjugate gradient method
are discussed.

SECTION 1
DEFINITION AND BASIC PROPERTIES

Suppose that H_1 and H_2 are Hilbert spaces over the same
field of scalars. We consider the fundamental problem of
solving a general linear equation of the type

(1) $Tx = b$

where $b \in H_2$ and $T \in L(H_1, H_2)$. The most prevalent example
of an equation of type (1) is the one which obtains when
$H_1 = R^n$, $H_2 = R^m$ and T is an m by n matrix. If $H_1 = H_2 = L^2[0,1]$, then the integral operator defined by

$$(Tx)(s) = \int_0^1 k(s,t)x(t)\,dt, \quad s \in [0,1]$$

where $k(s,t) \in L^2([0,1] \times [0,1])$ provides another important
example. If the operator T has an inverse then equation (1)
always has the unique solution $x = T^{-1}b$. But in general
such a linear equation may have more than one solution
($N(T) \neq \{0\}$) or may have no solution at all ($b \notin R(T)$). Even
if the equation has no solution in the traditional meaning it

is still possible to assign what is in a sense a "best possible" solution to the problem. In fact, if we let P denote the projection of H_2 onto $R(T)$ (assumed to be closed), then Pb is the vector in $R(T)$ which is closest to b and it seems reasonable to consider as a generalized solution of (1) any solution $u \in H_1$ of the equation

(2) $\qquad Tx = Pb.$

Let us consider for the sake of concreteness a simple two dimensional example. Suppose $T \in L(R^2, R^2)$ is represented in terms of the usual basis by the matrix

$$T = \begin{pmatrix} -1 & 1 \\ 1 & -1 \end{pmatrix}$$

and let

$$b = \begin{pmatrix} 2 \\ 1 \end{pmatrix}.$$

Then $R(T) = \text{span}\{(1,-1)\}$ and

$$Pb = \begin{pmatrix} 1/2 \\ -1/2 \end{pmatrix}$$

and the set of all generalized solutions is given by

$$\{(x_1, x_2) : x_2 = 1/2 + x_1\}.$$

CLOSED RANGE CASE 39

Another natural approach to assigning generalized solutions to equation (1) is to find a $u \in H_1$ which "comes closest" to solving (1) in the sense that

$$||Tu - b|| \leq ||Tx - b||$$

for any $x \in H_1$. It is geometrically evident that this is equivalent to the previous notion of generalized solution. In fact we may prove the following.

<u>Theorem 2.1.1.</u> Suppose $T \in L(H_1, H_2)$ has closed range and $b \in H_2$, then the following conditions on $u \in H_1$ are equivalent

(i) $\quad Tu = Pb$,
(ii) $\quad ||Tu - b|| \leq ||Tx - b||$ for any $x \in H_1$,
(iii) $\quad T^*Tu = T^*b$.

<u>Proof.</u>
(i) implies (ii): Suppose $Tu = Pb$. Then for any $x \in H_1$, we have by use of the Pythagorean property and the fact that $Pb - b \in R(T)^\perp$

$$\begin{aligned}||Tx - b||^2 &= ||Tx - Pb||^2 + ||Pb - b||^2 \\ &= ||Tx - Pb||^2 + ||Tu - b||^2 \\ &\geq ||Tu - b||^2.\end{aligned}$$

(ii) implies (iii): If $||Tu - b|| \leq ||Tx - b||$ for all $x \in H_1$, then again by use of the Pythagorean property and the fact that $Pb = Tx$ for some $x \in H_1$, we have

$$||Tu - b||^2 = ||Tu - Pb||^2 + ||b - Pb||^2$$
$$\geq ||Tu - Pb||^2 + ||b - Tu||^2.$$

Therefore

$$Tu - b = Pb - b \in R(T)^\perp = N(T^*)$$

and (iii) is satisfied.

(iii) implies (i): If (iii) holds, then $Tu - b \in R(T)^\perp$ and therefore

$$0 = P(Tu - b) = Tu - Pb. \qquad \#$$

Definition. A vector $u \in H_1$ which satisfies the equivalent conditions (i)-(iii) of (2.1.1) is called a <u>least squares solution</u> of the equation $Tx = b$.

Note that since $R(T)$ is closed a least squares solution of (1) exists for each $b \in H_2$. Also, if $N(T) \neq \{0\}$ then there are infinitely many least squares solutions of (1) since if u is a least squares solution then so is $u + v$ for any $v \in N(T)$.

CLOSED RANGE CASE   41

Ideally we would like to "invert" the operator T associating with each $b \in H_2$ some uniquely determined least squares solution $u \in H_1$. One natural way to do this is to note that by (2.1.1) the set of least squares solutions of (1) can be written as

$$\{u \in H_1 : T^*Tu = T^*b\}$$

which by the continuity and linearity of T and T* is a closed convex set. This set contains a unique vector of minimal norm by (1.1.4) and we will choose this vector to be the least squares solution uniquely associated with b by way of the generalized inversion process.

<u>Definition (V)</u>. Let $T \in L(H_1, H_2)$ have closed range. The mapping $T^\dagger : H_2 \to H_1$ defined by $T^\dagger b = u$, where u is the least squares solution of minimal norm of the equation $Tx = b$, is called the <u>generalized inverse</u> of T.

In the next section we will give several equivalent definitions of the generalized inverse and we will refer to definition (V) above as the <u>variational</u> definition. Note that if the operator T is invertible then we certainly have $T^\dagger = T^{-1}$.

<u>Theorem 2.1.2</u>. If $T \in L(H_1, H_2)$ has closed range, then $R(T^\dagger) = R(T^*) = R(T^\dagger T)$.

Proof. Let $b \in H_2$. We show first that $T^\dagger b \in N(T)^\perp = R(T^*)$. Suppose

$$T^\dagger b = u_1 + u_2 \in N(T)^\perp \oplus N(T),$$

then u_1 is a least squares solution of $Tx = b$ since

$$Tu_1 = T(u_1 + u_2) = TT^\dagger b = Pb.$$

Also, if $u_2 \neq 0$ we have by the Pythagorean property

$$||u_1||^2 < ||u_1 + u_2||^2 = ||T^\dagger b||^2$$

in contradiction to the fact that $T^\dagger b$ is the least squares solution of minimal norm. Therefore $T^\dagger b = u_1 \in N(T)^\perp$.

Conversely, suppose that $u \in N(T)^\perp$. Let $b = Tu$. We claim that $u = T^\dagger b$. We certainly have

$$Tu = PTu = Pb$$

therefore u is a least squares solution. If x is any other least squares solution, then

$$Tx = Pb = Tu$$

and hence $x - u = \bar{u} \in N(T)$. It follows that

CLOSED RANGE CASE 43

$$||x||^2 = ||u||^2 + ||\bar{u}||^2 \geq ||u||^2$$

hence u is the least squares solution of minimal norm, that is $u = T^{\dagger}b$. Thus we see that $R(T^{\dagger}) = R(T^*)$.

Note that for any $b \in H_2$,

$$T^{\dagger}b = T^{\dagger}Pb \in R(T^{\dagger}T)$$

and hence $R(T^{\dagger}) = R(T^{\dagger}T)$. #

<u>Corollary 2.1.3.</u> If $T \in L(H_1,H_2)$ has closed range, then $T^{\dagger} \in L(H_2,H_1)$.

<u>Proof.</u> Let $b,\bar{b} \in H_2$, then

$$TT^{\dagger}b = Pb \quad \text{and} \quad TT^{\dagger}\bar{b} = P\bar{b}.$$

Therefore

$$TT^{\dagger}b + TT^{\dagger}\bar{b} = P(b + \bar{b}) = TT^{\dagger}(b + \bar{b})$$

and hence by (2.1.2)

$$T^{\dagger}b + T^{\dagger}\bar{b} - T^{\dagger}(b + \bar{b}) \in N(T)^{\perp} \cap N(T) = \{0\}.$$

It can be shown in a similar way that for any scalar α

$$T^{\dagger}(\alpha b) = \alpha T^{\dagger}b.$$

Since $R(T^{\dagger}) = R(T^*) = N(T)^{\perp}$ we have by Theorem 1.2.3 the existence of a positive number m such that

$$||TT^{\dagger}b|| \geq m||T^{\dagger}b||$$

for all $b \in H_2$. Since $TT^{\dagger}b = Pb$ it follows that

$$||b|| \geq ||Pb|| \geq m||T^{\dagger}b||$$

and hence $T^{\dagger}$ is bounded. #

Since $R(T^{\dagger}) = N(T)^{\perp}$ and since $u + v$ is a least squares solution if u is a least squares solution and $v \in N(T)$ we see that the set of all least squares solutions of (1) has the form $T^{\dagger}b + N(T)$. The figure below gives an illustration of the situation for the two dimensional example considered above.

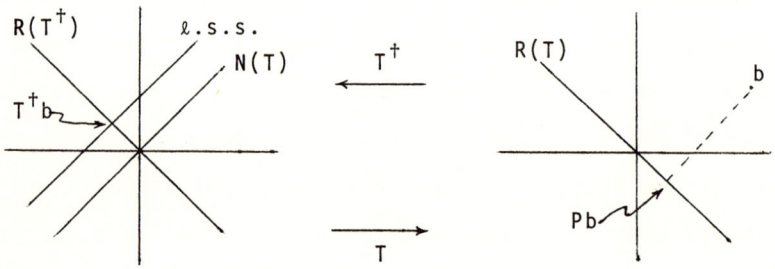

FIGURE 1

CLOSED RANGE CASE   45

It is easy to see that for this example

$$T^\dagger = \begin{pmatrix} -1/4 & 1/4 \\ 1/4 & -1/4 \end{pmatrix}.$$

Note that if $R(T)$ is closed, then $R(T*)$ is also closed by (1.2.4) and if we set $\tilde{T} = T*T|R(T*)$, it follows from (1.2.3) that

$$(\tilde{T}x,x) = ||Tx||^2 \geq m^2||x||^2 \qquad (m > 0)$$

for $x \in R(T*)$. Hence the inverse $\tilde{T}^{-1}$ can be defined on $R(\tilde{T})$. But it is easy to see that $R(\tilde{T}) = R(T*T) = R(T*)$, and hence by (1.2.2), $\tilde{T}^{-1} \in L(R(T*),R(T*))$. The next theorem now follows easily.

Theorem 2.1.4. Suppose $T \in L(H_1,H_2)$ has closed range and set $\tilde{T} = T*T|R(T*)$, then $T^\dagger = \tilde{T}^{-1}T*$.

The representation given in the next theorem is important from a computational viewpoint. It shows that if either of the spaces H_1 or H_2 is finite dimensional, then the computation of $T^\dagger$ where $T \in L(H_1,H_2)$ reduces to the computation of the generalized inverse of a matrix. For if H_1 is finite dimensional, then T^*T may be represented as a matrix while if H_2 is finite dimensional TT^* has a matrix representation.

Theorem 2.1.5. Suppose $T \in L(H_1, H_2)$ has closed range, then $T^\dagger = (T^*T)^\dagger T^* = T^*(TT^*)^\dagger$.

Proof. First we note that $R(T^*T) = R(T^*)$, for any $y \in H_2$ has a representation $y = y_1 + y_2 \in R(T) \oplus N(T^*)$. Hence $T^*y = T^*y_1 \in R(T^*T)$. Now if $b \in H_2$, then

$$T^*T(T^*T)^\dagger T^*b = P_{R(T^*T)} T^*b = T^*b$$

and hence $(T^*T)^\dagger T^*b$ is a solution of the normal equation (2.1.1) (iii), i.e., $(T^*T)^\dagger T^*b$ is a least squares solution. Therefore,

$$(T^*T)^\dagger T^*b = T^\dagger b + v$$

for some $v \in N(T)$. Since $(T^*T)^\dagger T^*b \in R((T^*T)^\dagger)$, we have by (2.1.2)

$$(T^*T)^\dagger T^*b \in R(T^*T) = R(T^*) = N(T)^\perp$$

and it follows that

$$(TT^*)^\dagger T^*b = T^\dagger b.$$

The proof of the other equality is similar.   #

SECTION 2
OTHER DEFINITIONS

Generalized inverses of matrices and linear operators may be defined in many different ways. We have chosen the variational definition given above because it is the most appealing definition to us in both its analytic and geometrical aspects. For other (not necessarily equivalent) definitions of generalized inverses the reader is referred to the books of Rao and Mitra [1] and Ben-Israel and Greville [1]. For a very interesting historical survey of generalized inverses see the paper of Reid [1].

E. H. Moore [1] was the first to give an explicit definition of the generalized inverse of an arbitrary matrix. This definition was given by Moore in an abstract published in the Bulletin of the American Mathematical Society in 1920 which was either little noticed or its significance was not realized (owing perhaps to Moore's very individualistic terminology and notation) for the concept remained undeveloped for decades thereafter. In our terminology Moore's definition is as follows.

<u>Definition (M)</u>. If $T \in L(H_1, H_2)$ has closed range, then $T^\dagger$ is the unique operator in $L(H_2, H_1)$ satisfying

$$TT^\dagger = P_{R(T)} \quad \text{and} \quad T^\dagger T = P_{R(T^\dagger)}.$$

In a series of important (and again little noticed) papers in 1949, Tseng [1], [2], [3] defined a generalized inverse for linear operators in Hilbert space (see the final chapter of Ben-Israel and Greville [1] for an exposition of Tseng's work). Bjerhammer [1] defined a generalized inverse for rectangular matrices in a paper published in 1951. The theory experienced a renaissance with the publication of a paper on matrix generalized inverses by Penrose [1] in 1955 and the literature of generalized inverses has proliferated rapidly ever since. Penrose was unaware of the work of Moore and Tseng when he published his paper; the equivalence of Penrose's definition of the generalized inverse to that of Moore was pointed out by Rado [1] in 1956. In our notation the Penrose definition may be phrased as follows.

Definition (P). If $T \in L(H_1, H_2)$ has closed range, then $T^\dagger$ is the unique operator in $L(H_2, H_1)$ satisfying

(1) $\quad TT^\dagger = (TT^\dagger)^*$,
(2) $\quad T^\dagger T = (T^\dagger T)^*$,
(3) $\quad TT^\dagger T = T$,

and

(4) $\quad T^\dagger TT^\dagger = T^\dagger$.

It is important to realize that this concept is well-defined. The next theorem shows that $T^\dagger$ is unique; the existence

CLOSED RANGE CASE   49

of $T^\dagger$ will follow when we establish the equivalence of the various definitions of generalized inverses.

Theorem 2.2.1. There can be at most one operator $T^\dagger \in L(H_2,H_1)$ satisfying definition (P).

Proof. (Penrose [1]).
  Note that by (4) and (1) we have

(5) $\quad T^\dagger = T^\dagger(TT^\dagger)^* = T^\dagger T^{\dagger*} T^*.$

It follows from (4) and (2) that

(6) $\quad T^\dagger = (T^\dagger T)^* T^\dagger = T^* T^{\dagger*} T^\dagger.$

By (3) and (1) we have

(7) $\quad T = (TT^\dagger)^* T = T^{\dagger*} T^* T$

and therefore

(8) $\quad T^* = T^* TT^\dagger.$

Finally, by (3) and (2) we see that

(9) $\quad T = TT^\dagger T = TT^* T^{\dagger*}$

and therefore

(10) $T^* = T^\dagger T T^*$.

Suppose now that X and Y are two operators in $L(H_2, H_1)$ satisfying definition (P), then

$$\begin{aligned} X &= XX^*T^* & \text{by (5)} \\ &= XX^*T^*TY & \text{by (8)} \\ &= XTY & \text{by (7)} \\ &= XTT^*Y^*Y & \text{by (6)} \\ &= T^*Y^*Y & \text{by (10)} \\ &= Y & \text{by (6).} \quad \# \end{aligned}$$

If $T^\dagger$ satisfies definition (M) then we see that $TT^\dagger$ and $T^\dagger T$ are self-adjoint (exercise 5, Chapter I). Also,

$$TT^\dagger T = P_{R(T)} T = T$$

and

$$T^\dagger T T^\dagger = P_{R(T^\dagger)} T^\dagger = T^\dagger,$$

hence $T^\dagger$ satisfies definition (P). Conversely, if $T^\dagger$ satisfies definition (P), then

$$(TT^\dagger)(TT^\dagger) = T(T^\dagger T T^\dagger) = TT^\dagger.$$

Therefore $TT^\dagger$ is a self-adjoint idempotent operator and

CLOSED RANGE CASE

hence is the projection operator onto the subspace
$S = \{y : TT^\dagger y = y\}$ (exercise 6, Chapter I). Since $TT^\dagger T = T$
it follows that $R(T) \subset S$. Also, if $y \in S$ then for any
$z \in N(T^*)$,

$$(y,z) = (TT^\dagger y, z) = (T^\dagger y, T^* z) = 0,$$

hence $S \subset N(T^*)^\perp = R(T)$. It follows that

$$TT^\dagger = P_{R(T)}.$$

We leave as an exercise the proof that if $T^\dagger$ satisfies definition (P), then

$$T^\dagger T = P_{R(T^\dagger)}.$$

Hence we see that definitions (M) and (P) are equivalent and we shall refer to these definition (either equivalent form) as the _Moore-Penrose_ definition, which we will denote by (M-P). We would like to show that the Moore-Penrose definition of $T^\dagger$ is equivalent to the variational definition given in the previous section, but first we will display yet another definition of $T^\dagger$ which was published by Desoer and Whalen [1] in 1963.

Definition (D-W). If $T \in L(H_1, H_2)$ has closed range, then $T^\dagger$ is the unique operator in $L(H_2, H_1)$ satisfying

(11) $\quad T^{\dagger}Tx = x \quad$ for $x \in N(T)^{\perp}$

and

(12) $\quad T^{\dagger}y = 0 \quad$ for $y \in R(T)^{\perp}$.

Theorem 2.2.2. The definition (V), (M-P) and (D-W) are equivalent.

Proof. (D-W) implies (V): Suppose $T^{\dagger}$ satisfies the (D-W) definition and let $b = b_1 + b_2 \in R(T) \oplus R(T)^{\perp} = H_2$. Then

$$TT^{\dagger}b = TT^{\dagger}b_1$$

and since $b_1 \in R(T)$ we have $b_1 = Tx$ for some $x \in N(T)^{\perp}$, therefore

$$TT^{\dagger}b = T(T^{\dagger}Tx) = Tx = b_1 = P_{R(T)}b ,$$

that is, $T^{\dagger}b$ is a least squares solution. Suppose that u is another least squares solution, then $u - T^{\dagger}b \in N(T)$. Also, since $b = b_1 + b_2 \in R(T) \oplus R(T)^{\perp}$ and since $T^{\dagger}b_2 = 0$ we have

$$T^{\dagger}b = T^{\dagger}b_1 = T^{\dagger}Tx = x$$

for some $x \in N(T)^{\perp}$. Therefore, $T^{\dagger}b \in N(T)^{\perp}$ and hence $T^{\dagger}b \perp (u - T^{\dagger}b)$. The Pythagorean property then gives $||T^{\dagger}b||^2 \leq ||u||^2$ and hence $T^{\dagger}$ satisfies definition (V).

CLOSED RANGE CASE 53

(V) implies (M-P): If $T^\dagger$ satisfies definition (V) then clearly $TT^\dagger = P$. Also for any $x \in H_1$ we have by (2.1.2)

$$x = x_1 + x_2 \in N(T)^\perp \oplus N(T) = R(T^\dagger) \oplus N(T).$$

Therefore

$$TT^\dagger Tx = TT^\dagger Tx_1 = PTx_1 = Tx_1$$

and hence

$$T^\dagger Tx - x_1 \in N(T) \cap N(T)^\perp,$$

that is,

$$T^\dagger Tx = x_1 = P_{R(T^\dagger)} x.$$

(M-P) implies (D-W): Suppose $T^\dagger$ satisfies definition (M-P) and $y \in R(T)^\perp$, then

$$T^\dagger y = T^\dagger TT^\dagger y = T^\dagger Py = 0.$$

It remains to show that if $T^\dagger$ satisfies definition (M-P), then $T^\dagger Tx = x$ for $x \in N(T)^\perp$. Since $TT^\dagger T = T$, it follows that $T^\dagger Tx - x \in N(T)$ for any $x \in H_1$. Now if $x \in N(T)^\perp$ we have by the Pythagorean property

$$\|x\|^2 \geq \|P_{R(T^\dagger)}x\|^2 = \|T^\dagger Tx\|^2$$

$$= \|T^\dagger Tx - x\|^2 + \|x\|^2.$$

Therefore $T^\dagger Tx = x$ for $x \in N(T)^\perp$ and hence $T^\dagger$ satisfies definition (D-W). #

SECTION 3
A REPRESENTATION THEOREM
AND APPLICATIONS

In this section we present a general representation theorem for the generalized inverse of a bounded linear operator with closed range along with a number of corollaries which result in specific representations and computational procedures. This representation theorem, which appears in Groetsch [2], may be viewed as an application of the classical theory of summability to the representation of generalized inverses. Summability theory has been used by Bellman [1] and Schönberg [1], [2] to represent solutions of integral equations, by Muller [1], [2], [3] to represent inverses of linear operators and by Mann [1] Niethammer and Schempp [1], Schempp [1], Reinermann [1], Koliha [1] and others to sum divergent iterative sequences.

The key to the representation theorem is Theorem 2.1.4

CLOSED RANGE CASE

along with the observation that if $A \in L(H,H)$ and $||I - A|| < 1$, then A is invertible and

$$A^{-1} = \sum_{k=0}^{\infty} (I - A)^k.$$

In order to motivate the representation theorem we make note of the following purely _formal_ identity

(1) $\quad (I_1 - (I_1 - T^*T))^{-1} T^* = T^{-1}$

where I_1 is the identity operator on H_1 and $T \in L(H_1, H_2)$. We shall show that certain general summability methods can be applied to the _formal_ series expansion

(2) $\quad (I_1 - (I_1 - T^*T))^{-1} = \sum_{k=0}^{\infty} (I_1 - T^*T)^k$

to construct $T^\dagger$ in analogy with formula (1). For our purposes we may consider as a "summability" method for the series

$$1 + (1 - x) + (1 - x)^2 + \cdots$$

any family $\{S_\beta(x)\}$ of continuous functions where the index β runs through a directed set (i.e. a "net" of continuous functions) with the property that $\lim_\beta S_\beta(x) = x^{-1}$ uniformly on a suitable region.

Suppose now that $T \in L(H_1, H_2)$ has closed range and let $H = R(T^*)$. Then H is a Hilbert space and the spectrum of the operator $\tilde{T} \in L(H,H)$ defined by

$$\tilde{T} = T^*T|H$$

satisfies $\sigma(\tilde{T}) \subset (0, \infty)$. Indeed, by (1.2.3) and the fact that $R(T^*) = N(T)^\perp$, we have for each $x \in H$

$$((T^*T|H)x,x) = ||Tx||^2 \geq m^2||x||^2$$

where m is some positive number (the norm above is the norm on H inherited from H_1). Therefore $T^*T|H$ is a self-adjoint positive operator on H and the relation $\sigma(\tilde{T}) \subset (0, \infty)$ follows.

<u>Theorem 2.3.1</u>. Suppose $T \in L(H_1, H_2)$ has closed range and let $\tilde{T} = T^*T|H$, where $H = R(T^*)$. If Ω is an open set with $\sigma(\tilde{T}) \subset \Omega \subset (0, \infty)$ and $\{S_\beta(x)\}$ is a family of continuous real valued functions on Ω with $\lim_\beta S_\beta(x) = 1/x$ uniformly on $\sigma(\tilde{T})$, then

$$T^\dagger = \lim_\beta S_\beta(\tilde{T})T^*$$

where the convergence is in the uniform topology for $L(H_2, H_1)$. Furthermore,

CLOSED RANGE CASE 57

$$||S_\beta(\tilde{T})T^* - T^\dagger|| \leq \sup_{x \in \sigma(\tilde{T})} |xS_\beta(x) - 1|\ ||T^\dagger||.$$

<u>Proof</u>. By use of the spectral theorem for self-adjoint linear operators (Chapter I, section 3) we have

$$\lim_\beta S_\beta(\tilde{T}) = \tilde{T}^{-1}$$

uniformly in L(H,H). It follows by (2.1.4) that

$$\lim_\beta S_\beta(\tilde{T})T^* = \tilde{T}^{-1}T^* = T^\dagger.$$

To obtain the error bound we note that $T^* = \tilde{T}T^\dagger$ and therefore

$$S_\beta(\tilde{T})T^* - T^\dagger = (S_\beta(\tilde{T})\tilde{T} - I)T^\dagger.$$

Since $\tilde{T}$ is self-adjoint and $S_\beta(x)x$ is a real-valued continuous function on $\sigma(\tilde{T})$, it follows that $S_\beta(\tilde{T})\tilde{T}$ is self-adjoint (such is evidently the case for polynomials with real coefficients and $S_\beta(x)x$ can be uniformly approximated by such polynomials according to the Wierstrass Theorem). The spectral radius formula for self-adjoint operators (see e.g. Lorch [1, p. 109]) now gives

$$||S_\beta(\tilde{T})\tilde{T} - I|| = |\sigma(S_\beta(\tilde{T})\tilde{T} - I)|.$$

But $\sigma(S_\beta(\tilde{T})\tilde{T} - I) = \{S_\beta(x)x - 1 : x \in \sigma(\tilde{T})\}$ follows from the spectral mapping theorem, again because $S_\beta(x)x - 1$ can be uniformly approximated on $\sigma(\tilde{T})$ by polynomials. Therefore we see that

$$||S_\beta(\tilde{T})\tilde{T} - I|| = \sup_{x \in \sigma(\tilde{T})} |S_\beta(x)x - 1|$$

and hence

$$||S_\beta(\tilde{T})T^* - T^\dagger|| \leq \sup_{x \in \sigma(\tilde{T})} |S_\beta(x)x - 1| \, ||T^\dagger||. \quad \#$$

In using the error estimate above on specific approximation procedures it will be convenient to have a lower bound for $\sigma(\tilde{T})$. This is provided by the next theorem.

Theorem 2.3.2. Suppose $T \in L(H_1, H_2)$ has closed range and let $\tilde{T} = T^*T|H$, where $H = R(T^*)$. Then $\lambda \geq ||T^\dagger||^{-2}$ for each $\lambda \in \sigma(\tilde{T})$.

Proof. If $x \in H$, then

$$||x||^2 = ||P_{R(T^*)}x||^2 = ||T^\dagger Tx||^2$$
$$\leq ||T^\dagger||^2 ||Tx||^2 = ||T^\dagger||^2 (\tilde{T}x, x).$$

Therefore $(\tilde{T}x, x) \geq ||T^\dagger||^{-2} ||x||^2$, that is $\tilde{T} \geq ||T^\dagger||^{-2} I$. The result now follows by Theorem 1.3.1. $\quad \#$

CLOSED RANGE CASE

We now present several examples which illustrate the use of the above results in developing specific representations and computational procedures for $T^\dagger$ and corresponding error bounds.

Example 1: __An Operator-Valued Integral Representation__.

One way to produce a family of functions $\{S_\beta(x)\}$ which is suitable for use in the theorem above is to employ the classical Borel summability transform (see Hardy [1]) on the geometric series

$$1 + (1 - x) + (1 - x)^2 + \cdots.$$

An infinite series $\sum_{n=0}^{\infty} a_n$ is said to be Borel summable to the value a if

$$\lim_{t \to \infty} S_t = a$$

where

(3) $$S_t = \int_0^t e^{-u} \sum_{n=0}^{\infty} \frac{a_n u^n}{n!} du.$$

The value of the Borel transform (or any summability transform) resides in the fact that if $\sum_{n=0}^{\infty} a_n$ converges, then $\{S_t\}$

converges to the same limit, however, the transforms (3) may converge even if the original series diverges and hence summability transforms allow one to associate generalized sums to certain divergent series. The following classical theorem of Borel (for a proof see Hardy [1]) shows that these transforms may be used to analytically extend a power series $f(z) = \sum_{n=0}^{\infty} a_n(1-z)^n$ beyond its circle of convergence within the "Borel polygon" of $f(z)$.

<u>Theorem</u> (Borel). Let $f(z) = \sum_{n=0}^{\infty} a_n(1-z)^n$ be regular at 1 and suppose f has singularities $z_1,\ldots,z_m$. Let L_i be the line through z_i perpendicular to the vector $\vec{1z_i}$ and let H_i be the half-plane containing 1 determined by this line. Then the series $\sum_{n=0}^{\infty} a_n(1-z)^n$ is Borel summable to $f(z)$ (uniformly on compact subsets) inside of the <u>Borel polygon</u> $\bigcap_{i=1}^{m} H_i$.

Of course, we have no use for Borel's theorem in its full generality. For our purposes it is sufficient to consider the function

$$f(x) = 1/x.$$

Note that the Borel polygon of f contains $(0,\infty)$ and the Borel transform of the geometric series expansion of f is

CLOSED RANGE CASE 61

$$S_t(x) = \int_0^t e^{-xu}\, du.$$

In this special case the proof of (2.3.2) becomes trivial:

$$\int_0^\infty e^{-xu}\, du = \lim_{t\to\infty} (1 - e^{-tx})/x$$
$$= 1/x.$$

uniformly on compact subsets of $(0,\infty)$. Therefore we may apply Theorem (2.3.1) to obtain:

<u>Corollary 2.3.3</u>. If $T \in L(H_1, H_2)$ has closed range, then

$$T^\dagger = \int_0^\infty e^{-T^*Tu} T^*\, du$$

where the operator valued integral converges in the uniform topology of $L(H_2, H_1)$.

This representation, except for a change of variables, is the integral representation proved by Showalter [1] using a different method (see also the final chapter of Ben-Israel and Greville [1]). Showalter [1] also gave the following error bound

$$||T^\dagger - T^\dagger(t)|| \leq ||T^\dagger|| \exp(-t\, ||T^\dagger||^{-2})$$

where

$$T^{\dagger}(t) = \int_0^t e^{-T^*Tu} T^* du.$$

This follows directly from (2.3.1) and (2.3.2), for we see that

$$S_t(x)x - 1 = x \int_0^t e^{-xu} du - 1 = -e^{-xt}$$

and hence for $x \in \sigma(\tilde{T})$ we have by (2.3.2)

$$|S_t(x)x - 1| \leq \exp(-||T^{\dagger}||^{-2}t)$$

and the error bound follows by (2.3.1).

Example 2: <u>An Iterative Method</u>.

Another well known summability method is the Euler-Knopp method (see Knopp [1] and Hardy [1]). A series $\sum_{n=0}^{\infty} a_n$ is said to be Euler-Knopp summable with parameter α to the value a if the sequence defined by

$$S_n = \alpha \sum_{k=0}^{n} \sum_{j=0}^{k} \binom{k}{j} (1-\alpha)^{k-j} \alpha^j a_j$$

converges to a. If $a_k = (1-x)^k$ for $k = 0,1,2,\ldots$, then we obtain as the Euler-Knopp transform of the series $\sum_{k=0}^{\infty} (1-x)^k$, the sequence given by

CLOSED RANGE CASE 63

$$S_n(x) = \alpha \sum_{k=0}^{n} [1 - \alpha x]^k.$$

Clearly

$$\lim_{n\to\infty} S_n(x) = 1/x$$

uniformly on compact subsets of the set

$$E_\alpha = \{x : |1 - \alpha x| < 1\}$$
$$= \{x : 0 < x < 2/\alpha\}.$$

Since on the Hilbert space $H = R(T^*)$ we have

$$0 < T^*T|H \leq ||T||^2 I,$$

it follows by (1.3.1) that

$$\sigma(\tilde{T}) \subset (0, ||T||^2]$$

and hence we may apply Theorem (2.3.1) if we choose the parameter α in such a way that

$$(0, ||T||^2] \subset E_\alpha.$$

This is surely the case if $0 < \alpha < 2||T||^{-2}$, hence for such α we have the representation

$$T^\dagger = \alpha \sum_{k=0}^{\infty} [I_1 - \alpha T^*T]^k T^*.$$

Note that if we set

(4) $$T_n^\dagger = \alpha \sum_{k=0}^{n} [I_1 - \alpha T^*T]^k T^*$$

then

(5) $$T_0^\dagger = \alpha T^*, \quad T_{n+1}^\dagger = (I_1 - \alpha T^*T)T_n^\dagger + \alpha T^*.$$

Therefore we have proved the following

<u>Corollary 2.3.4</u>. If $T \in L(H_1, H_2)$ has closed range, then the sequence $\{T_n^\dagger\}$ defined by (5), where $0 < \alpha < ||T||^{-2}$, converges to $T^\dagger$ in the uniform topology for $L(H_2, H_1)$.

This iterative sequence has also been studied by Showalter [1], Petryshyn [2] and others (see Altman [1] for invertible operators and Bialy [1] for a pointwise version of (5)).

There are several ways of regarding the representation (5). We have taken the point of view of Euler-Knopp summability above, but in view in (2.1.4) we could attempt to approximate $T^\dagger$ by operators of the form

$$p_n(\tilde{T})T^*$$

CLOSED RANGE CASE

where $\tilde{T} = T^*T|H$ and $p_n(x)$ is a suitable polynomial approximation to x^{-1}. Now if we expand the function $h(x) = x^{-1}$ in a Taylor series about the point α^{-1}, then we have

$$(6) \qquad h(x) = \alpha - \alpha^2(x - \alpha^{-1}) + \alpha^3(x - \alpha^{-1})^2 + \cdots$$

$$= \alpha \sum_{k=0}^{\infty} (1 - \alpha x)^k$$

where the series converges uniformly on closed subintervals of $0 < x < 2/\alpha$. Representing the operator $\tilde{T}$ by x, this condition translates to $0 < \alpha < 2||T||^{-2}$, and therefore we see from (6) that the representation (4) may be regarded as an expansion of the operator $\tilde{T}^{-1}T^*$ in a Taylor series about the operator $\alpha^{-1}I_1$.

The iterative method (5) may be considered in yet another light. Recall that by (2.1.1) the vector $T^\dagger b$ is a minimizer of the functional f on H_1 defined by

$$f(x) = (Tx - b, Tx - b).$$

Following Rosenbloom's [1] discussion of the continuous steepest descent method we consider x as a function $x(t)$ of the real variable t with $x(0) = 0$ and note that the path $x(t)$ of most rapid decrease of the functional f satisfies

(7) $$\frac{dx}{dt} = -\nabla f(x) = -T^*(Tx - b)$$

(see the next section for a more complete discussion of the idea of steepest descent). Given a first order ordinary differential equation

$$\frac{dx}{dt} = F(x,t), \quad x(0) = x_0,$$

then perhaps the simplest method for approximating $x(t)$ is to choose a discretization parameter $\alpha > 0$ and use the Euler method (see Conte and deBoor [1]) to generate the approximations $\{x_n\}$ given by

$$x_{n+1} = x_n + \alpha F(x_n, t_n)$$

where $t_n = x_0 + n\alpha$. Using this method on the differential equation (7) above, we have

$$\begin{aligned} x_{n+1} &= x_n + \alpha(-T^*(Tx_n - b)) \\ &= (I_1 - \alpha T^*T)x_n + \alpha T^*b, \end{aligned}$$

which the reader will immediately recognize as a pointwise version of (5) where

$$x_n = T_n^\dagger b.$$

CLOSED RANGE CASE

Finally, we mention that the iterative process (5) arises from the method of successive approximations for computing fixed points. In fact, we may regard 1/x as a fixed point of the function

$$S(y) = (1 - \alpha x)y + \alpha, \qquad \alpha > 0.$$

In order to approximate this fixed point we may use the sequence of successive approximations (see e.g. Conte and deBoor [1]) defined by

(8) $\qquad S_0(x) = \alpha, \quad S_{n+1}(x) = (1 - \alpha x)S_n(x) + \alpha.$

It is a simple matter to verify that

$$\lim_n S_n(x) = 1/x$$

uniformly on compact subsets of $\{x : 0 < x < 2/\alpha\}$. Hence if $0 < \alpha < 2||T||^{-2}$, then we may apply Theorem (2.3.1) to show that

$$\lim_n S_n(\tilde{T})T^* = T^{\dagger}$$

where $\tilde{T} = T^*T|H$. But it is easy to see from (8) that $S_n(\tilde{T})T^* = T_n^{\dagger}$, where $T_n^{\dagger}$ is given by (5).

68 GENERALIZED INVERSES

Finally, we derive an error estimate for this iterative process. We suppose that $0 < \alpha < 2||T||^{-2}$.

If $S_n(x)$ is defined as in (8), then

$$S_{n+1}(x)x - 1 = (1 - \alpha x)[S_n(x)x - 1]$$

therefore, since $S_0(x) = \alpha$,

$$|S_n(x)x - 1| = |1 - \alpha x|^{n+1}.$$

Since $||T^\dagger||^{-2} \leq x \leq ||T||^2$ for $x \in \sigma(\tilde{T})$ and $0 < \alpha < 2||T||^{-2}$, it follows that

$$|1 - \alpha x| \leq \beta$$

where β is given by

(9) $\qquad \beta = \max\{|1 - \alpha||T||^2|, |1 - \alpha||T^\dagger||^{-2}|\}.$

Note that since

$$||T|| \; ||T^\dagger|| \geq ||TT^\dagger|| = ||P|| = 1$$

we have

$$2 > \alpha||T||^2 \geq \alpha||T^\dagger||^{-2} > 0$$

CLOSED RANGE CASE   69

and therefore $0 < \beta < 1$. From (2.3.1) we now obtain the error estimate

$$||T_n^+ - T^+|| \le ||T^+|| \beta^{n+1}.$$

Example 3: <u>An Analog of Schulz's Method</u>.

To develop another iterative method we regard $1/x$ as the root of the function

$$S(y) = y^{-1} - x.$$

The Newton-Raphson method (see Conte and deBoor [1]) can be used to approximate this root. This is done by generating a sequence y_n where

$$\begin{aligned} y_{n+1} &= y_n - S(y_n)/S'(y_n) \\ &= y_n(2 - x\, y_n), \end{aligned}$$

for suitable y_0. Suppose that for $\alpha > 0$ we define a sequence of functions $\{S_n(x)\}$ by

(10) $S_0(x) = \alpha, S_{n+1}(x) = S_n(x)(2 - x\, S_n(x))$.

<u>Lemma 2.3.5</u>. The sequence of functions defined by (10) converges to x^{-1} uniformly on compact subsets of $E_\alpha = \{x : 0 < x < 2/\alpha\}$.

Proof. Note that

$$xS_{n+1}(x) - 1 = -(xS_n(x) - 1)^2.$$

Iterating on this equality, it follows that if x is confined to a compact subset of E_α, then there is a constant K (depending on the compact set) with $0 < K < 1$ and

$$|xS_n(x) - 1| = |x\alpha - 1|^{2^n} \leq K^{2^n},$$

and the lemma follows. #

The great attraction of the Newton-Raphson method is the generally quadratic nature of the convergence, which is displayed in the proof of (2.3.5). Using this lemma in conjunction with (2.3.1) we see that the sequence of operators $\{S_n(\tilde{T})\} \subset L(H,H)$, where $H = R(T^*)$, defined by

$$S_0(\tilde{T}) = \alpha I, \quad S_{n+1}(\tilde{T}) = S_n(\tilde{T})(2I - T^*TS_n(\tilde{T}))$$

has the property that $\lim_n S_n(\tilde{T})T^* = T^\dagger$, uniformly in $L(H_2,H_1)$. Note that if we set $T_n = S_n(\tilde{T})T^*$, then

(11) $\quad T_0 = \alpha T^* \quad$ and $\quad T_{n+1} = T_n(2I_2 - TT_n)$

where I_2 is the identity operator on H_2.

CLOSED RANGE CASE

Corollary 2.3.6. If $T \in L(H_1,H_2)$ has closed range, then the sequence $\{T_n\}$ defined by (11), where $0 < \alpha < 2||T||^{-2}$, converges to $T^\dagger$ in the uniform topology for $L(H_2,H_1)$.

The iterative process above was apparently first studied by Schulz [1] in 1933 for nonsingular matrices (see also Altman [2], Albrecht [1], Ansorge [1] and Dück [1]).

If $x \in \sigma(\tilde{T})$ and $0 < \alpha < 2||T||^{-2}$, then we have seen that $|1 - \alpha x| < \beta$ where β is given by (9) above. It follows as in the proof of (2.3.5) that

$$|xS_n(x) - 1| \leq \beta^{2^n}$$

and hence from (2.3.1) we obtain the error bound

$$||T_n - T^\dagger|| \leq ||T^\dagger||\beta^{2^n},$$

(see also Showalter [1]).

Example 4: Hyperpower Methods.

The so-called hyperpower methods are a device for extrapolating on the desirable quadratic convergence property of Schulz's method to obtain iterative methods which display pth order convergence, where $p \geq 2$ is an integer. Of course in order to gain a more favorable order of convergence a price is paid in that more computation is required per iteration. For discussions of the hyperpower method for matrices and linear operators see Altman [2], John [1], Lonseth [1] and Petryshyn [1].

Given an integer $p \geq 2$, define functions $\{S_n^p(x)\}$ by

$$S_0^p(x) = \alpha > 0, \quad S_{n+1}^p(x) = S_n^p(x) \sum_{k=0}^{p-1} (1 - xS_n^p(x))^k.$$

If $p = 2$, these formulas agree with those given in (10), but note that

$$(12) \qquad |xS_{n+1}^p(x) - 1| = |xS_n^p(x) - 1|^p$$

which indicates the pth order convergence of the process. It follows easily from (12) that if x lies in a compact subset of $E_\alpha = \{x : 0 < x < 2/\alpha\}$, then there is a constant K with $0 < K < 1$ such that

$$|xS_n^p(x) - 1| = |x\alpha - 1|^{p^n} \leq K^{p^n}$$

and hence

$$\lim_n S_n^p(x) = 1/x$$

CLOSED RANGE CASE 73

uniformly on compact subsets of E_α. Theorem (2.3.1) may therefore be applied to the sequence of operators $\{S_n^p(\tilde{T})T^*\}$ where

$$S_0^p(\tilde{T}) = \alpha I, \quad S_{n+1}^p(\tilde{T}) = S_n^p(\tilde{T}) \sum_{k=0}^{p-1} (I - T^*TS_n^p(\tilde{T}))^k.$$

Note that

$$S_{n+1}^p(\tilde{T})T^* = S_n^p(\tilde{T}) \sum_{k=0}^{p-1} (I - T^*TS_n^p(\tilde{T}))^k T^*$$

$$= S_n^p(\tilde{T})T^* \sum_{k=0}^{p-1} (I_2 - TS_n^p(\tilde{T})T^*)^k$$

and hence if $\{T_n^p\} \subset L(H_2, H_1)$ is defined by

(13) $\quad T_0^p = \alpha T^*, \quad T_{n+1}^p = T_n^p \sum_{k=0}^{p-1} (I_2 - TT_n^p)^k,$

then since $\sigma(\tilde{T}) \subset E_\alpha$ for $0 < \alpha < 2||T||^{-2}$, we may conclude the following.

<u>Corollary 2.3.7</u>. If $T \in L(H_1, H_2)$ has closed range, then the sequence $\{T_n^p\} \subset L(H_2, H_1)$ defined by (13), where $0 < \alpha < 2||T||^{-2}$, converges to $T^\dagger$ in the uniform topology for $L(H_2, H_1)$.

If β is given by (9) above, then it follows from (12) that

$$|S_n^p(x)x - 1| \leq \beta^{p^n}$$

for $x \in \sigma(\tilde{T})$. We therefore obtain from (2.3.1)

$$||T^\dagger - T_n^p|| \leq ||T^\dagger||\beta^{p^n}.$$

Example 5. **Tihonov Regularization**.

According to the variational definition, $T^\dagger b$ is the vector $u \in H_1$ which minimizes the functional $||Tx - b||$ and also has smallest norm among all such minimizing vectors. The idea of Tihonov's regularization of order zero (see Tihonov [1]) is to approximately minimize both the functional $||Tx - b||$ and the norm $||x||$ by minimizing the functional $g : H_1 \to R$ defined

(14) $$g(x) = ||Tx - b||^2 + t||x||^2$$

where $t > 0$. The minimum of this functional will occur at the unique stationary point u of g, i.e. the vector u which satisfies $\nabla g(u) = 0$. Now by exercise 14 of Chapter I, the gradient of g is given by

$$\nabla g(x) = 2(T^*Tx - T^*b) + 2tx$$

and hence the unique minimizer u_t of (14) satisfies

CLOSED RANGE CASE 75

$$u_t = (tI_1 + T^*T)^{-1}T^*b.$$

On intuitive grounds it seems reasonable to expect that

$$\lim_{t \to 0^+} u_t = T^\dagger b.$$

In fact a much stronger result holds, for if we set

$$S_t(x) = (t + x)^{-1} \quad (t > 0)$$

in Theorem 2.3.1, we obtain the following.

<u>Corollary 2.3.8</u>. If $T \in L(H_1,H_2)$ has closed range, then

$$T^\dagger = \lim_{t \to 0^+} (tI_1 + T^*T)^{-1}T^*$$

uniformly in $L(H_2,H_1)$.

This representation was apparently first given for square matrices by den Broeder and Charnes (see Ben-Israel and Charnes [1]). To obtain an error bound for this method note that for $x \in \sigma(\tilde{T})$ we have

$$|xS_t(x) - 1| = \frac{t}{x + t} \leq \frac{t}{||T^\dagger||^{-2} + t}.$$

Therefore from (2.3.1) we conclude that

$$||(tI_1 + T^*T)^{-1}T^* - T^\dagger|| \leq \frac{t||T^\dagger||^3}{1 + t||T^\dagger||^2}.$$

Example 6: <u>A Method Based on Interpolatory Function Theory</u>.

All of the examples we have considered so far were based on approximating the function 1/x and then representing these approximations via the spectral theorem. J. C. Dunn [1] has shown that the inverse of an invertible normal operator can be calculated by interpolating the function x^{-1} at certain integral values and considering the corresponding sequence of polynomials in the operator as approximations to the inverse. In a similar vein we will approximate the generalized inverse by interpolating the function f(x) = 1/x and using Theorem (2.3.1). By use of the Newton-Gregory interpolation formula (see Conte and deBoor [1]), the unique polynomial of degree $\leq$ n which interpolates f(x) at x = 1,2,...,n+1, may be written as

$$p_n(x) = \sum_{k=0}^{n} \binom{x-1}{k} \Delta^k f(1)$$

where Δ is the forward difference operator defined by

$$\Delta f(x) = f(x+1) - f(x), \quad \Delta^k f(x) = \Delta(\Delta^{k-1}f)(x)$$

CLOSED RANGE CASE 77

and

$$\binom{x}{0} = 1,$$

$$\binom{x}{k} = \frac{(x)(x-1)\cdots(x-k+1)}{k!}.$$

It is not hard to see that if $f(x) = 1/x$, then $\Delta^k f(1) = (-1)^k/(k+1)$. A routine calculation now shows that the interpolating polynomial is given by

$$p_n(x) = \sum_{k=0}^{n} \frac{1}{k+1} \prod_{j=0}^{k-1} \left(1 - \frac{x}{1+j}\right).$$

<u>Lemma 2.3.9.</u> $1 - xp_n(x) = \prod_{j=0}^{n} \left(1 - \frac{x}{1+j}\right).$

<u>Proof.</u> The statement is clearly true for $n = 0$. Also, assuming that it is true for n, we have

$$1 - xp_{n+1}(x) = 1 - xp_n(x) - \frac{x}{n+2} \prod_{j=0}^{n} \left(1 - \frac{x}{1+j}\right)$$

$$= \left(1 - \frac{x}{n+2}\right) \prod_{j=0}^{n} \left(1 - \frac{x}{1+j}\right)$$

$$= \prod_{j=0}^{n+1} \left(1 - \frac{x}{1+j}\right). \quad \#$$

<u>Lemma 2.3.10.</u> The polynomials $p_n(x)$ satisfy $\lim_n p_n(x) = 1/x$, uniformly on compact subsets of $(0,\infty)$.

Proof. If $x > 0$, then by (2.3.9)

$$1 - xp_n(x) = \prod_{j=0}^{n} \left(1 - \frac{x}{1+j}\right)$$

and hence it suffices to show that

(15) $$\lim_n \prod_{j=0}^{n} \left(1 - \frac{x}{1+j}\right) = 0$$

uniformly on compact subsets of $(0,\infty)$. If x lies in a fixed compact subset of $(0,\infty)$, then there is a constant $K > 0$ such that

$$0 < 1 - \frac{x}{1+j} < 1 - \frac{K}{1+j},$$

therefore to establish (15) it suffices to show that

$$\prod_{j=0}^{\infty} \left(1 - \frac{K}{1+j}\right) = 0.$$

But this is a consequence of a well-known fact about infinite products. Namely, if $\{a_j\}$ is a sequence of numbers with $0 < a_j < 1$ and $\sum_{j=0}^{\infty} a_j = \infty$, then the sequence $\prod_{j=0}^{n} (1 - a_j)$ is nonincreasing and positive and therefore has a limit $a \geq 0$. It is easy to show by induction that

$$a_0 + \sum_{k=0}^{n} a_{k+1} \prod_{j=0}^{k} (1 - a_j) = 1 - \prod_{j=0}^{n+1} (1 - a_j)$$

CLOSED RANGE CASE   79

and hence

$$1 \geq 1 - \prod_{j=0}^{n+1} (1 - a_j)$$

$$= a_0 + \sum_{k=0}^{n} a_{k+1} \prod_{j=0}^{k} (1 - a_j)$$

$$\geq a \sum_{k=0}^{n} a_k.$$

But since $\sum_{j=0}^{\infty} a_j = \infty$, it follows that $a = 0$. #

The lemmas above allow us to conclude that

$$T^{\dagger} = \lim_{n} p_n(\tilde{T})T^*$$

where $\tilde{T} = T^*T|R(T^*)$. But in order to phrase this result in a form which is more convenient for computation we note that $p_0(x) = 1$ and

$$p_{n+1}(x) = p_n(x) + \frac{1}{n+2} \prod_{j=0}^{n} (1 - \frac{x}{1+j})$$

$$= p_n(x) + \frac{1}{n+2}[1 - xp_n(x)]$$

by (2.3.9). Therefore if we set

$$T_n = p_n(\tilde{T})T^*$$

we see that

(16) $\quad T_0 = T^*, \quad T_{n+1} = T_n + \dfrac{1}{n+2}[T^* - T^*TT_n].$

Hence we may state

<u>Corollary 2.3.11.</u> If $T \in L(H_1, H_2)$ has closed range, then the sequence $\{T_n\} \subset L(H_2, H_1)$ defined by (16) converges to $T^\dagger$ in the uniform topology for $L(H_2, H_1)$.

The representation of $T^\dagger$ given in this corollary was presented by Groetsch [2] where it was motivated in terms of Lototsky transforms (see e.g. Agnew [1]).

To obtain an asymptotic error bound for this method, note that for

$$x \in \sigma(\tilde{T}) \subset [||T^\dagger||^{-2}, ||T||^2]$$

and for $j \geq J = [||T||^2]$, we have

$$1 - \dfrac{x}{1+j} \leq \exp(-x/(1+j))$$

and therefore

$$\prod_{j=J}^{n} \left(1 - \dfrac{x}{1+j}\right) \leq \exp\left(-x \sum_{j=J}^{n} \dfrac{1}{1+j}\right).$$

CLOSED RANGE CASE

Also,

$$\sum_{j=J}^{n} \frac{1}{1+j} \geq \int_{J+1}^{n+2} \frac{dt}{t} = \log(n+2) - \log(J+1)$$

and hence

$$\exp(-x \sum_{j=J}^{n} \frac{1}{1+j}) \leq (J+1)^x (n+2)^{-x} = ||T||^{2x}(n+2)^{-x}.$$

Therefore if we set

$$C = \max_{x \in \sigma(\tilde{T})} \left| ||T||^{2x} \prod_{j=0}^{J-1} (1 - \frac{x}{1+j}) \right|,$$

then we have by (2.3.9)

$$|1 - xp_n(x)| \leq C(n+2)^{-x}$$

and using (2.3.1) we obtain

$$||T_n - T^\dagger|| \leq C||T^\dagger||(n+2)^{-||T^\dagger||^{-2}}$$

for n sufficiently large.

Example 7: <u>Other representations</u>.

Theorem 2.3.1 can be used to develop representations for $T^\dagger$ by using for $S_\beta(x)$ any summability transform which sums the geometric series

$$1 + (1 - x) + (1 - x)^2 + \cdots$$

to $1/x$ uniformly on compact subsets of $(0,\infty)$. The classical summability transforms of Mittag-Leffler, Lindelöf and LeRoy (see Hardy [1]) enjoy this property and they may be used to obtain the following representations of $T^\dagger$ (see Groetsch [2]).

<u>Corollary 2.3.12.</u> If $T \in L(H_1,H_2)$ has closed range then the following representations of $T^\dagger$ are convergent in the uniform topology for $L(H_2,H_1)$:

$$T^\dagger = \lim_{t \to 0^+} \sum_{k=0}^{\infty} \frac{1}{\Gamma(1 + tk)} [I_1 - T^*T]^k T^*$$

$$T^\dagger = T^* + \lim_{t \to 0^+} \sum_{k=1}^{\infty} e^{-tk\log k}[I_1 - T^*T]^k T^*$$

$$T^\dagger = \lim_{t \to 0^+} \sum_{k=0}^{\infty} \frac{\Gamma(1 + (1 - t)k)}{\Gamma(1 + k)}[I_1 - T^*T]^k T^*.$$

SECTION 4

<u>STEEPEST DESCENT</u>

In 1847 Cauchy [1] proposed what has since become known as the method of steepest descent for solving systems of

CLOSED RANGE CASE   83

nonlinear equations. His idea was to interpret the solution
of a system of nonlinear equations as a minimizer of a certain
nonnegative functional. He then constructed an iterative
sequence in such a way that in moving from one approximation
to the next he proceeded in the direction of most rapid decrease of the functional.

In this section we will present some results of Nashed [1]
on approximating $T^\dagger b$, where $T \in L(H_1, H_2)$ has closed range
and $b \in H_2$, by use of the steepest descent method. Before
applying the method to singular linear operator equations we
will first give a brief description of the general method.

Suppose H is a Hilbert space and that $f : H \to R$ is a nonnegative functional. We seek a point $x^* \in H$ such that

$$f(x^*) = \inf\{f(x) : x \in H\}.$$

We assume that f is Fréchet differentiable at each point
of H. Given an initial approximation x_0 we try to minimize
f by moving in the direction of most rapid decrease of f.
We must then choose a direction $z \in H$ such that the directional derivative $Df(x_0, z)$ (see Chapter I, section 4) is as
small as possible. Now

$$Df(x_0, z) = (z, \nabla f(x_0))$$

and by Schwarz's inequality

$$-\|z\| \|\nabla f(x_0)\| \leq (z, \nabla f(x_0))$$

with equality only if z is a positive multiple of $-\nabla f(x_0)$. It follows that starting at x_0 the direction of most rapid decrease of f is in the direction of $-\nabla f(x_0)$. Hence we form a new approximation x_1 by starting at x_0 and taking a step $\alpha_0 > 0$ in the direction $-\nabla f(x_0)$, i.e.

$$x_1 = x_0 - \alpha_0 \nabla f(x_0).$$

The parameter α_0 is chosen in such a way that x_1 gives a minimum for f on the ray emanating from x_0 in the direction $-\nabla f(x_0)$, therefore α_0 must satisfy

$$\frac{d}{d\alpha} f(x_0 - \alpha \nabla f(x_0))\Big|_{\alpha=\alpha_0} = 0.$$

The iterative sequence $\{x_n\}$ is then generated by

(1) $$x_{n+1} = x_n - \alpha_n \nabla f(x_n)$$

where

(2) $$\frac{d}{d\alpha} f(x_n - \alpha \nabla f(x_n))\Big|_{\alpha=\alpha_n} = 0.$$

Suppose now that $T \in L(H_1, H_2)$ has closed range. Since least squares solutions of $Tx = b$ are just the minima of the functional

CLOSED RANGE CASE   85

$$f(x) = \tfrac{1}{2}||Tx - b||^2,$$

we shall approximate least squares solutions by applying the steepest descent method to f. By exercise 14 of Chapter I, the gradient of f is given by

$$\nabla f(x) = T^*Tx - T^*b$$

and by exercise 15 of this chapter, the optimal value of α is given by

$$\alpha = ||r||^2/||Tr||^2$$

where $r = T^*Tx - T^*b.$

Therefore we will consider the sequence $\{x_n\} \subset H_1$ given by

(3)
$$\begin{aligned} x_{n+1} &= x_n - \alpha_n r_n \\ r_n &= T^*Tx_n - T^*b \\ \alpha_n &= ||r_n||^2/||Tr_n||^2. \end{aligned}$$

Note that if $r_n = 0$, then x_n is a least squares solution and the method terminates at x_n. Also, if $Tr_n = 0$ then $r_n = 0$ since $r_n \in R(T^*) = N(T)^\perp$. Hence we will assume that $r_n \neq 0$ for all n.

Lemma 2.4.1. If $T \in L(H_1, H_2)$, then $\lim_n r_n = 0$. ($R(T)$ need not be closed.)

Proof. (Nashed [1]). Note that

$$f(x_{n+1}) = \tfrac{1}{2}||Tx_n - \alpha_n Tr_n - b||^2$$

$$= f(x_n) - \tfrac{1}{2}||r_n||^4/||Tr_n||^2.$$

Using this formula recursively, we obtain

$$f(x_{n+1}) = f(x_0) - \tfrac{1}{2} \sum_{i=0}^{n} ||r_i||^4/||Tr_i||^2.$$

Since $f(x) \geq 0$ for all x, it follows that

$$||T||^{-2} \sum_{i=0}^{\infty} ||r_i||^2 \leq \sum_{i=0}^{\infty} ||r_i||^4/||Tr_i||^2 \leq 2f(x_0),$$

therefore $||r_i|| \to 0$. #

Theorem 2.4.2. If $T \in L(H_1, H_2)$ has closed range then the steepest descent sequence converges to a least squares solution of $Tx = b$ for any $x_0 \in H_1$. The sequence converges to $T^\dagger b$ if and only if $x_0 \in R(T^*)$.

Proof. (Nashed [1]). Note that the steepest descent sequence satisfies

CLOSED RANGE CASE 87

$$x_{n+1} = x_0 - \sum_{i=0}^{n} \alpha_i r_i.$$

We will show that $\{x_n\}$ is a Cauchy sequence. For $m > n$ we have

$$x_m - x_n = -\sum_{i=n}^{m-1} \alpha_i r_i \in R(T^*).$$

By (1.2.3) there is a positive number δ such that

$$\delta^2 ||x_m - x_n||^2 \leq (T^*T(x_m-x_n), x_m-x_n) = ||T(x_m-x_n)||^2.$$

Therefore

$$(T^*T(x_m-x_n), x_m-x_n) \leq [||T^*Tx_m - T^*b|| + ||T^*Tx_n - T^*b||] \cdot ||x_m-x_n||$$

$$\leq \delta^{-1}[||T^*Tx_m-T^*b||+||T^*Tx_n-T^*b||] \cdot$$

$$||T(x_m-x_n)||.$$

Since $r_i \to 0$, we have

$$r_m - r_n = T^*T(x_m-x_n) \to 0$$

as $m,n \to \infty$.

By (1.2.2) T^* has a bounded inverse on $R(T) = N(T^*)^\perp$

and hence $\{T(x_m - x_n)\}$ is bounded, say $||T(x_m - x_n)|| \leq M$, therefore

$$\delta^2 ||x_m - x_n||^2 \leq \frac{M}{\delta}[||T^*Tx_m - T^*b|| + ||T^*Tx_n - T^*b||]$$

$$= M(r_m + r_n)/\delta \to 0 \quad \text{as } m, n \to \infty.$$

Hence $\{x_n\}$ is a Cauchy sequence and therefore converges to a vector $u \in H_1$. Since

$$T^*Tx_n - T^*b = r_n \to 0$$

we see that

$$T^*Tu = T^*b,$$

that is, u is a least squares solution.

Finally, since the set of all least squares solutions is given by $T^\dagger b \oplus N(T)$ and since $R(T^*) = N(T)^\perp$, we see that $T^\dagger b$ is the only least squares solution in $R(T^*)$. If $x_0 \in R(T^*)$, then

$$x_n = x_0 - \sum_{i=0}^{m-1} \alpha_i r_i \in R(T^*)$$

and since $R(T^*)$ is closed, it follows that $u \in R(T^*)$. But since u is a least squares solution we must then have $u = T^\dagger b$.

CLOSED RANGE CASE

If $x_0 \notin R(T^*)$, then $x_0 = x_0' + P_{N(T)}x_0$ where $x_0' \in N(T)^\perp = R(T^*)$ and $P_{N(T)}x_0 \neq 0$. Since $T^*TP_{N(T)}x_0 = 0$, we then have

$$x_n = x_0' - \sum_{i=0}^{n-1}\alpha_i r_i + P_{N(T)}x_0$$

and since $x_0' \in R(T^*)$,

$$x_0' - \sum_{i=0}^{n-1}\alpha_i r_i \to T^\dagger b, \text{ as } n \to \infty.$$

Therefore

$$x_n \to T^\dagger b + P_{N(T)}x_0, \text{ as } n \to \infty.$$

This completes the proof. #

Nashed [1] has given the following error bound for the steepest descent method:

$$||T^\dagger b + P_{N(T)}x_0 - x_n|| \leq C\left[\frac{M-m}{M+m}\right]^n$$

where C is a constant and m and M satisfy

$$m||x||^2 \leq (T^*Tx, x) \leq M||x||^2$$

for all $x \in R(T^*)$.

SECTION 5

THE CONJUGATE GRADIENT METHOD

In this section we will investigate the conjugate gradient method of Hestenes and Stiefel [1] (see also Hayes [1], Antosiewicz and Rheinboldt [1] and Daniel [1]) for approximating least squares solutions and generalized inverses. Lucid expositions of the geometry of the method in finite dimensional spaces may be found in Stiefel [1] and Faddeev and Faddeeva [1].

Suppose that T is an invertible matrix, then the unique solution u of the equation Tx = b is also the unique minimizer of the quadratic functional f defined by

$$f(x) = ||Tx - b||^2.$$

Note that

$$f(x) = (Ax,x) - 2(x,T^*b) + ||b||^2$$

where $A = T^*T$ is positive definite. The level sets of the functional f, that is the sets given by

(1) $\{x : f(x) = \text{constant}\}$,

are ellipsoids centered on the solution u. In fact, for a

CLOSED RANGE CASE

suitable orthogonal matrix V we have

$$A = VDV^*$$

where D is a positive diagonal matrix (see e.g. Forsythe and Moler [1]). The level sets (1) therefore take the form

(2) $\quad (Dw,w) - (2w,\tilde{b}) = \text{constant}$

where $w = V^*x$ and $\tilde{b} = V^*T^*b$. Now the set of points w which satisfy (2) is clearly an ellipsoid with center $D^{-1}\tilde{b} = V^*u$. If v is a vector which is tangent to the ellipsoid given by (2) at the point w, then

$$(Dw - \tilde{b}, v) = 0,$$

or equivalently,

$$(w - D^{-1}\tilde{b}, Dv) = 0.$$

But this says that if q is a vector emanating from a point w on the ellipsoid (2) and directed toward the center $D^{-1}\tilde{b}$ of the ellipsoid, then q and v are <u>conjugate directions</u> relative to the ellipsoid, i.e,

$$(q, Dv) = 0.$$

Expressing this in terms of the original coordinate system ($q \to Vp$, $v \to V\bar{p}$) we see that a vector p which is tangent to the ellipsoid (1) at the point x_0 and a vector $\bar{p}$ which emanates from x_0 and is directed toward the center of the ellipsoid (1) satisfy the <u>conjugacy relation</u>

$$(p, A\bar{p}) = 0.$$

Suppose now that $T \in L(H,H)$ is an invertible operator on the Hilbert space H. We now give a brief discussion (following Hayes [1]) of the <u>conjugate direction</u> method for solving $Tx = b$. Suppose that $\{p_i\}$ is a sequence of conjugate directions for the operator $A = T^*T$ (note that $A \geq mI$ for some $m > 0$), that is

(3) $$(p_i, Ap_j) = 0, \quad \text{for } i \neq j.$$

We choose an arbitrary starting point x_0 and move in the direction of p_0 by a step α_0,

$$x_1 = x_0 + \alpha_0 p_0.$$

The unique solution u of $Tx = b$ is also the unique minimizer of the quadratic functional

CLOSED RANGE CASE

(4) $$f(x) = ||Tx - b||^2$$
$$= (Tx - Tu, Tx - Tu)$$
$$= (A(x - u), x - u)$$
$$= (Ax,x) - 2(x,y) + (u,y)$$

where $A = T^*T$ and

$$y = Au = T^*b.$$

The step size α_0 is to be chosen optimally, i.e. so that f is minimized at the point x_1 on the ray emanating from x_0 in the direction p_0, that is

$$\frac{d}{d\alpha} f(x_0 + \alpha p_0)\Big|_{\alpha=\alpha_0} = 0.$$

By use of (4) and a routine computation we see that this is equivalent to

$$(p_0, A(x_0 + \alpha_0 p_0) - y) = 0$$

or

$$\alpha_0 = (p_0, r_0)/(p_0, Ap_0)$$

where $r_0 = y - Ax_0$. In general the iteration is given by

(5) $$x_{n+1} = x_n + \alpha_n p_n$$

where

(6) $$r_n = y - Ax_n$$

and

(7) $$\alpha_n = (p_n, r_n)/(p_n, Ap_n).$$

Lemma 2.5.1. The vectors $\{r_n\}$ defined by (6) satisfy the following:

(a) $(p_i, r_j) = 0$, for $i < j$,
(b) $(p_i, r_i) = (p_i, r_{i-1}) = \cdots = (p_i, r_1)$,
(c) $\alpha_i = (p_i, r_1)/(p_i, Ap_i).$

Proof. Note that

(8) $$r_{n+1} = y - Ax_{n+1}$$
$$= r_n - \alpha_n Ap_n.$$

Therefore for $i = 0, 1, 2, \ldots$, we have

$$(p_i, r_{i+1}) = (p_i, r_i) - \alpha_i(p_i, Ap_i) = 0,$$

by (7). Also, if $(p_i, r_j) = 0$ for some $j > i$, then

$$(p_i, r_{j+1}) = (p_i, r_j) - \alpha_j(p_i, Ap_j)$$
$$= -\alpha_j(p_i, Ap_j) = 0,$$

CLOSED RANGE CASE   95

by (3). Hence (a) is proved by induction. By (8) and (3) we have

$$(p_i, r_i) = (p_i, r_{i-1}) - \alpha_{i-1}(p_i, Ap_{i-1})$$
$$= (p_i, r_{i-1})$$

and property (b) follows. Property (c) follows immediately from (b) and (7). #

Lemma 2.5.2. The vector x_n minimizes the functional (4) on the set $\{x_0 + z : z \in \text{span}\{p_0, p_1, \ldots, p_{n-1}\}\}$.

Proof. (Hayes [1]). Suppose

$$x = x_0 + a_0 p_0 + a_1 p_1 + \cdots + a_{n-1} p_{n-1}$$

for some scalars $a_0, a_1, \ldots, a_{n-1}$. Then by (5) and (2.5.1) we have

$$(9) \quad (r_n, x - x_n) = \sum_{i=0}^{n-1}(a_i - \alpha_i)(r_n, p_i) = 0.$$

It follows that the functional f defined by (4) satisfies

$$(10) \quad f(x) - f(x_n) = (Ax, x) - (Ax_n, x_n) - 2(x - x_n, y)$$
$$= (Ax, x) - (Ax_n, x_n) - 2(x - x_n, r_n + Ax_n)$$
$$= (A, x,) - (Ax_n, x) - (x, Ax_n) + (x_n, Ax_n)$$
$$= (A(x - x_n), x - x_n),$$

96   GENERALIZED INVERSES

where we have used (9). Since A is positive definite we
obtain $f(x) > f(x_n)$ for $x \neq x_n$. #

We now make an additional assumption about the space H.
If we let

$$E_n = \text{span}\{p_0, \ldots, p_{n-1}\}$$

where $\{p_i\}$ are vectors satisfying (3), we shall assume that
the Hilbert space satisfies the following condition:

(P) $$H = \overline{\bigcup_{n=0}^{\infty} E_n}.$$

Such vectors $\{p_n\}$ can always be found if H is separable (a mild
assumption from the point of view of applications), i.e.,
if H contains a countable dense subset. For then we can find
a dense sequence of linearly independent vectors in H and
orthogonalize it relative to the inner product $\langle \cdot, \cdot \rangle$ given by

$$\langle x, y \rangle = (Ax, y),$$

to obtain a sequence $\{p_i\}$ satisfying (3). Hestenes [1] has
also given a general method of generating conjugate directions.

<u>Theorem 2.5.3</u>. Suppose H_1 is a Hilbert space satisfying (P)
and $T \in L(H_1, H_2)$ has a bounded inverse. Then the sequence $\{x_n\}$

CLOSED RANGE CASE   97

defined by (5) converges to the unique solution u of Tx = b for any $x_0 \in H_1$.

Proof. (Hayes [1]). By (2.5.2) $x_n - x_0$ minimizes f on the space E_n. Since $H_1 = \bigcup_{n=0}^{\infty} E_n$, there is a sequence of vectors $\{z_n\}$ such that $z_n \in E_n$ and $\lim_n z_n = w$, where w is the unique solution of

$$Aw = A(u - x_0).$$

We therefore have

$$f(z_n) \geq f(x_n - x_0) \geq 0$$

and $\lim_n f(z_n) = f(w)$ by the continuity of f. Hence

$$\lim_n f(x_n - x_0) = f(w).$$

It follows as in (10) that

$$f(z_n) - f(x_n - x_0) = (A(z_n - x_n + x_0), z_n - x_n + x_0)$$
$$\geq m||z_n - x_n + x_0||^2$$

where $m > 0$ satisfies $A \geq mI$. Hence we must have

$$w = \lim_n z_n = \lim_n (x_n - x_0).$$

But, since $Aw = Au - Ax_0$, this implies that $\lim_n x_n = u$. #

In summary, to find the unique solution u of $Ax = y$, where A is a positive definite operator on a Hilbert space H, we interpret u as the center of the infinite dimensional "ellipsoids" given by

(11) $\qquad (Ax,x) - 2(x,y) = $ constant.

Given conjugate directions $\{p_i\}$ we generate a sequence of approximations $\{x_n\}$ by (5). Since $\{p_i\}$ are conjugate directions and the step sizes α_i are chosen optimally we see that x_n is the center of the n-dimensional ellipsoid which is the intersection of the ellipsoid (11) with the set

$$\{x_0 + z : z \in \text{span}\{p_0, p_1, \ldots, p_{n-1}\}\}.$$

Thus, assuming condition (P), the centers of these ellipsoids with ever increasing dimensions converge to the center $A^{-1}y$ of the infinite dimensional ellipsoid (11).

Since the vectors $\{p_i\}$ are linearly independent we see that if the space H has finite dimension N, then the conjugate direction method converges in at most N steps.

CLOSED RANGE CASE

In the conjugate direction method the nonzero vectors $\{p_i\}$ are chosen arbitrarily save for the restriction that they satisfy (3). In fact all of the conjugate directions may be generated even before the approximations $\{x_n\}$ are computed. In the _conjugate gradient_ method a sequence of conjugate directions is generated one by one at each step of the iterative process. Before considering the conjugate gradient method for a singular linear operator with closed range we will first concern ourselves with nonsingular operators.

We preface our discussion of the conjugate gradient method by a brief reconsideration of the steepest descent method.

Note that if $x(t)$ is a level curve for the functional f, i.e.

$$f(x(t)) = c$$

for all $t \in [0,1]$ where c is a constant, then differentiating with respect to t and using (1.4.2) we obtain

$$f'(x(t))x'(t) = (x'(t), \nabla f(x(t))) = 0.$$

Therefore given a point x_0, the steepest descent direction $-\nabla f(x_0) = -r_0$ is always normal to the level curve of f through x_0. Since the normals to an ellipse do not in general pass through the center of the ellipse, it would seem reasonable, after taking one steepest descent step to obtain

x_1, to then move in the direction p_1 of the center of the elliptical level curve which passes through the point x_1 and lies in the plane determined by the vectors r_0 and r_1, instead of proceeding in the direction of steepest descent, viz. $-r_1$. The vector p_1 must then satisfy the conjugacy relation

$$(x_1 - x_0, Ap_1) = 0$$

or

$$(r_0, Ap_1) = 0,$$

where $A = T^*T$.

The idea of the conjugate gradient method therefore is as follows. First choose $x_0 \in H$ arbitrarily and set

(12) $$p_0 = r_0 = T^*Tx_0 - T^*b$$

and

(13) $$\alpha_0 = ||r_0||^2 / ||Tr_0||^2.$$

Now take a steepest descent step and compute

(14) $$x_1 = x_0 - \alpha_0 p_0,$$

then let

CLOSED RANGE CASE

$$r_1 = T^*Tx_1 - T^*b$$
$$= r_0 - \alpha T^*Tp_0$$

and choose a vector p_1 in the plane of r_1 and p_0 which is A-conjugate to p_0. Therefore

$$p_1 = r_1 + \beta_0 p_0$$

where the constant β_0 is chosen so that

$$(p_0, Ap_1) = 0,$$

i.e.
$$\beta_0 = -(p_0, Ar_1)/(p_0, Ap_0)$$
$$= -(r_1, Ap_0)/(p_0, Ap_0).$$

Note that this amounts to projecting r_1 onto the subspace which is orthogonal relative to the inner product $<x,y> = (x, Ay)$ to the subspace spanned by $\{p_0\}$. Now we set

$$x_2 = x_1 - \alpha_1 p_1$$

where α_1 is chosen optimally, that is

$$\frac{d}{d\alpha} f(x_1 - \alpha p_1)\Big|_{\alpha=\alpha_1} = 0.$$

This gives

$$\alpha_1 = (r_1, p_1)/||Tp_1||^2.$$

Continuing in this way, the conjugate gradient method for minimizing $f(x) = ||Tx - b||^2$ is generated by the following prescription:

$x_0 \in H$ is arbitrary,
$r_0 = p_0$ is given by (12),
α_0 is given by (13),
x_1 is given by (14),

and for $i = 1, 2, \ldots$, we compute

$$(15) \quad r_i = T^*Tx_i - T^*b = r_{i-1} - \alpha_{i-1} T^*Tp_{i-1}$$

where

$$(16) \quad \alpha_{i-1} = (r_{i-1}, p_{i-1})/||Tp_{i-1}||^2.$$

If $r_i \neq 0$, we compute

$$(17) \quad p_i = r_i + \beta_{i-1} p_{i-1}$$

where

CLOSED RANGE CASE

(18) $$\beta_{i=1} = -(r_i, T^*Tp_{i-1})/||Tp_{i-1}||^2.$$

Finally, we set

(19) $$x_{i+1} = x_i - \alpha_i p_i.$$

<u>Lemma 2.5.4</u>. The vectors $\{p_i\}$ defined by (17) are A-conjugate and the vectors $\{r_i\}$ defined by (15) are orthogonal, furthermore, $(r_k, r_k) = (r_k, p_k)$ for $k = 0, 1, \cdots$.

<u>Proof</u>. (John [1]). We certainly have

$$(p_0, Ap_1) = 0 = (Ap_0, p_1)$$

and

$$\begin{aligned}(r_0, r_1) &= (p_0, p_0 - \alpha_0 Ap_0) \\ &= (p_0, p_0) - \alpha_0(p_0, Ap_0) \\ &= 0.\end{aligned}$$

Suppose that

$$(p_i, Ap_j) = 0 = (r_i, r_j)$$

for $i \neq j$ and $i \leq k$, $j \leq k$. We will show that

$$(r_i, r_{k+1}) = 0 \text{ and } (p_i, Ap_{k+1}) = 0,$$

104   GENERALIZED INVERSES

if $i \leq k$ and the lemma will follow by induction. Now by (15) and (17),

$$(20) \quad (r_i, r_{k+1}) = (r_i, r_k) - \alpha_k(r_i, Ap_k)$$
$$= (r_i, r_k) - \alpha_k(p_i, Ap_k) + \alpha_k \beta_{i-1}(p_{i-1}, Ap_k)$$
$$= 0$$

for $i < k$, by the induction hypothesis. If $i = k$, we have

$$(r_i, r_{k+1}) = (r_k, r_k) - \alpha_k(p_k, Ap_k)$$
$$= (r_k, r_k) - (r_k, p_k)$$
$$= -\beta_{k-1}(r_k, p_{k-1})$$

by (20), (16), and (17). By use of (15) and (16) we then have

$$(r_i, r_{k+1}) = -\beta_{k-1}(r_{k-1} - \alpha_{k-1} Ap_{k-1}, p_{k-1})$$
$$= -\beta_{k-1}[(r_{k-1}, p_{k-1}) - \alpha_{k-1}(Ap_{k-1}, p_{k-1})]$$
$$= 0.$$

Note that we have also shown that

$$(21) \quad (r_k, r_k) = (r_k, p_k), \quad k = 0, 1, \cdots.$$

Now for $i < k$, we have by (17) and (15)

CLOSED RANGE CASE   105

$$(Ap_i, p_{k+1}) = (p_i, Ar_{k+1}) + \beta_k(p_i, Ap_k)$$
$$= (p_i, Ar_{k+1}) = (Ap_i, r_{k+1})$$
$$= \alpha_i^{-1}(r_i - r_{i+1}, r_{k+1}) = 0.$$

(Note that we may assume that $\alpha_i \neq 0$ since if $\alpha_i = 0$, then by (16) and (21), $r_i = 0$ and the iteration is terminated at the previous step.) Finally, if $i = k$, then by (17) and (18)

$$(p_k, Ap_{k+1}) = (p_k, Ar_{k+1} + \beta_k Ap_k) = 0$$

which completes the proof. #

The next theorem follows immediately from (2.5.3) and (2.5.4).

<u>Theorem 2.5.5</u>. Suppose H_1 is a Hilbert space satisfying (P) and $T \in L(H_1, H_2)$ is invertible. Then the conjugate gradient method converges to the unique solution u of $Tx = b$ for any $x_0 \in H_1$.

We now present a result which gives additional geometrical insight into the conjugate gradient method and also shows that the convergence of the method is monotonic.

<u>Theorem 2.5.6</u>. Let C_n be the closed convex hull of

$\{x_0, x_1, \ldots, x_n\}$. Then x_n is the unique vector in C_n which is closest to the solution u of $Tx = b$.

Proof. Suppose $x = \sum_{i=0}^{n} \lambda_i x_i$, where $0 \le \lambda_i$ and $\sum_{i=0}^{n} \lambda_i = 1$. Then $x_n - x = \sum_{i=0}^{n} \overline{\lambda}_i x_i$, where $\overline{\lambda}_i = -\lambda_i$ for $i < n$ and $\overline{\lambda}_n = 1 - \lambda_n$. Since $x_i = x_0 - \sum_{j=0}^{i-1} \alpha_j p_j$, for $i \ge 1$, it follows that

$$x_n - x = (\sum_{i=0}^{n-1} \overline{\lambda}_i) x_0 + (1 - \lambda_n) x_0 + \sum_{k=0}^{n-1} \gamma_k p_k$$

$$= \sum_{k=0}^{n-1} \gamma_k p_k$$

where $\gamma_k = -\sum_{j=0}^{k} \lambda_j \alpha_k$. Now by (16), (21) and the fact that A is positive definite we see that $\alpha_k \ge 0$ and hence $\gamma_k \le 0$. If u is the solution of $Tx = b$, then by (2.5.5)

$$u - x_n = -\sum_{j=n}^{\infty} \alpha_j p_j$$

and hence

$$(u - x_n, x_n - x) = -\sum_{k=0}^{n-1} \sum_{j=n}^{\infty} \alpha_j \gamma_k (p_j, p_k).$$

Now by exercise (17) it follows that

CLOSED RANGE CASE   107

$$(u - x_n, x_n - x) \geq 0.$$

An appeal to exercise (2) of Chapter I completes the proof. #

Daniel [2] has given the following error estimate for the conjugate gradient method. If u is the unique solution of Tx = b, then

$$(22) \quad ||x_n - u||^2 \leq \left(\frac{M - m}{M + m}\right)^{2n} ||Tx_0 - b||^2 / m$$

where m and M are positive numbers such that $mI \leq T^*T \leq MI$.

Suppose now that $T \in L(H_1, H_2)$ has closed range and we wish to compute $T^\dagger b$ for a given $b \in H_2$. In the following theorem of Kammerer and Nashed [2] the conjugate gradient method is used to approximate $T^\dagger b$. Since R(T) is closed, so is $R(T^*)$ by (1.2.4). Below we will denote the Hilbert space $R(T^*)$ by H and we assume that H satisfies (P) in order to apply the conjugate gradient method. The functional g appearing below is defined by

$$g(x) = ||Tx - Pb||^2$$

where P is the projection of H_2 onto R(T).

<u>Theorem 2.5.7</u>. If $T \in L(H_1, H_2)$ has closed range, then the

conjugate gradient method converges monotonically to the least squares solution $u = T^{\dagger}b + (I_1 - P_H)x_0$ of the equation $Tx = b$, where P_H is the projection of H_1 onto the closed subspace $H = R(T^*)$. Moreover, if m and M are positive numbers such that

$$mI \leq T^*T|H \leq MI$$

where I is the identity on H, then

(23) $$\|x_n - u\|^2 \leq \frac{g(x_0)}{m} \left(\frac{M-m}{M+m}\right)^{2n}.$$

Proof. (Kammerer and Nashed [2]). Note that for any $x_0 \in H_1$, $\{r_i\} \subset H$ and $\{p_i\} \subset H$, therefore $\{x_i\} \subset x_0 + H$. Also the mapping $R : x_0 + H \to H$ obtained by restricting P_H to $x_0 + H$ is an isometry (i.e. $\|Rx\| = \|x\|$) onto H and the conjugate gradient method applied to the operator $S \in L(H,R(T))$ defined by $S = T|H$ generates a sequence $\{x_n'\}$ which is related to $\{x_n\}$ by $x_n' = Rx_n$. Also S has a bounded inverse and therefore (2.5.5) and (2.5.6) apply to the sequence $\{x_n'\}$, showing that $\{x_n'\}$ converges monotonically to $T^{\dagger}b$, the unique solution of $Sx = Pb$. Daniel's error bound (22) gives

$$\|x_n' - T^{\dagger}b\|^2 \leq \frac{g(x_0)}{m} \left(\frac{M-m}{M+m}\right)^{2n}.$$

By applying the isometry R^{-1} we see that

CLOSED RANGE CASE 109

$$R^{-1}T^{\dagger}b = T^{\dagger}b + (I_1 - P_H)x_0 = u$$

and $R^{-1}x_n' = x_n$. Therefore $\{x_n\}$ converges monotonically to u and the error bound (23) holds. #

Note that if $x_0 \in R(T^*)$, then $(I_1 - P_H)x_0 = 0$ and in this case the conjugate gradient method converges monotonically to $T^{\dagger}b$.

The conjugate gradient method has the important property that in the finite dimensional case it converges in a finite number of steps. In fact, if $T \in L(H_1,H_2)$ has rank r (i.e. R(T) is r-dimensional), then $H = R(T^*)$ is r-dimensional and since $\{p_i\} \subset H$ it follows from (2.5.2) that the sequence $\{x_n'\}$ will terminate in at most $k \leq r$ steps at a value x_k' satisfying

$$||Sx_k' - Pb|| = 0.$$

But this implies that $x_k' = T^{\dagger}b$ and therefore $\{x_n\}$ terminates at the kth step where

$$x_k = R^{-1}x_k' = T^{\dagger}b + (I_1 - P_H)x_0.$$

This proves the following result of Kammerer and Nashed [2].

<u>Corollary 2.5.8.</u> If $T \in L(H_1,H_2)$ has rank r, then for any $x_0 \in H_1$ the conjugate gradient method for $Tx = b$ converges in at most r steps to the least squares solution $T^{\dagger}b + (I_1-P_H)x_0$.

EXERCISES

1. Show that if $T \in L(R^n, R^n)$ is a diagonal matrix, $T = \text{diag}(a_1, a_2, \ldots, a_n)$, then $T^\dagger = \text{diag}(\hat{a}_1, \hat{a}_2, \ldots, \hat{a}_n)$ where

$$\hat{a}_i = \begin{cases} 0 & \text{if } a_i = 0 \\ 1/a_i & \text{if } a_i \neq 0 \end{cases}$$

2. Let $T \in L(R^n, R^n)$ be a real n-by-n matrix. Then there is a diagonal matrix D and two orthogonal matrices U and V such that $T = UDV^*$ (see e.g. Forsythe and Moler [1]). Show that $T^\dagger = VD^\dagger U^*$, where $D^\dagger$ may be calculated as in exercise 1.

3. If $T \in L(H_1, H_2)$ has closed range, show that $N(T) = R(I_1 - T^\dagger T)$.

4. Show that if $P \in L(H, H)$ is a projection operator, then $P^\dagger = P$.

5. Show by example that the relation $(TS)^\dagger = S^\dagger T^\dagger$ does not hold in general.

6. Compute $T^\dagger$ if T is the linear operator represented by the matrix $\begin{pmatrix} 1 & 0 & 1 \\ 0 & 2 & 2 \end{pmatrix}$.

CLOSED RANGE CASE 111

7. Show that if $T \in L(H_1, H_2)$ has closed range, then $T^{\dagger\dagger} = T$.

8. Show that if $T \in L(R^n, R)$ is represented by $T = (t_1, \ldots, t_n)$, then $T^{\dagger} = (\sum_{i=1}^{n} t_i^2)^{-1} T^*$.

9. Show that if $T \in L(H_1, H_2)$ has closed range, then $R(T) = R(TT^{\dagger})$.

10. Show that if $T \in L(H, H)$ has closed range and is self-adjoint, then $TT^{\dagger} = T^{\dagger}T$.

11. Show that if $T \in L(H_1, H_2)$ has closed range, then $T^{\dagger *} = T^{* \dagger}$.

12. Show that if $T^{\dagger}$ satisfies definition (P), then $T^{\dagger}T = P_{R(T^{\dagger})}$.

13. Show that if $T^{\dagger}$ satisfies (M - P), then $R(T^{\dagger}) = N(T)^{\perp}$.

14. Show that if $T \in L(H_1, H_2)$ has closed range and $\hat{T} = TT^*|R(T)$, then $T^{\dagger} = T^* \hat{T}^{-1}$.

15. Let $f(x) = \frac{1}{2} ||Tx - b||^2$ where $T \in L(H_1, H_2)$ and $b \in H_2$. Show that the unique number $\bar{\alpha}$ satisfying

$$\frac{d}{d\alpha} f(x - \alpha \nabla f(x))\Big|_{\alpha = \bar{\alpha}} = 0$$

is given by $\bar{\alpha} = ||r||^2 / ||Tr||^2$, where

$$r = T^*Tx - T^*b.$$

16. Show that if $A \in L(H,H)$ is symmetric and positive definite, then the unique solution u of $Ax = y$ is also the unique minimum of the functional defined by (4), Section 5. Also, verify that the optimal value of α in the conjugate direction method is given by
$$\alpha_0 = (p_0, r_0)/(p_0, Ap_0).$$

17. Show that the vectors $\{p_j\}$ in the conjugate gradient method satisfy $(p_j, p_k) = ||r_j||^2 ||p_k||^2 / ||r_k||^2 \geq 0$, for $k \leq j$.

CHAPTER III
GENERALIZED INVERSES OF BOUNDED LINEAR OPERATORS WITH ARBITRARY RANGE

In this chapter the notion of the generalized inverse of a bounded linear operator is extended to the important case of operators whose ranges are not necessarily closed. Equations involving operators of this type occur frequently in applications, indeed any integral operator whose kernel does not have finite rank provides an example of an operator with nonclosed range. A fundamental distinction between the case of an operator with closed range and the case of an operator with nonclosed range is that the generalized inverse of an operator with nonclosed range turns out to be an unbounded operator and therefore approximations to such a generalized inverse by bounded operators must necessarily converge only in the pointwise sense at best.

SECTION 1
DEFINITION AND BASIC PROPERTIES

If $T \in L(H_1, H_2)$ and $R(T)$ is not necessarily closed, then we cannot always define the projection P of H_2 onto $R(T)$ as in Section 1 of the previous chapter. Of course, in trying to solve

(1) $\qquad Tx = b$

where $b \in H_2$ is not necessarily in $R(T)$ we may define the operator Q to be the projection of H_2 onto $\overline{R(T)}$ and assign as generalized solutions of (1) any solution u of the equation

$$(2) \qquad Tx = Qb.$$

But the situation now is much different from in the previous chapter as equation (2) may not have a solution, i.e. it may happen that $Qb \notin R(T)$. A modification of the argument in Chapter II proves the following:

Theorem 3.1.1. Let $T \in L(H_1, H_2)$ and let Q be the projection of H_2 onto $\overline{R(T)}$, then the following conditions on $u \in H_1$ are equivalent

(i) $\quad Tu = Qb$,
(ii) $\quad ||Tu - b|| \leq ||Tx - b||$ for any $x \in H_1$,
(iii) $\quad T^*Tu = T^*b$.

Proof. The proofs that (i) implies (ii) and (iii) implies (i) are the same as in Theorem 2.1.1 with P replaced by Q (note that $R(T)^\perp = \overline{R(T)}^\perp$). To see that (ii) implies (iii), observe that since $Qb \in \overline{R(T)}$ there is a sequence $\{x_n\}$ in H_1 such that $Qb = \lim_n Tx_n$ and hence

… ARBITRARY RANGE CASE

$$||b - Qb||^2 = \lim_n ||b - Tx_n||^2 \geq ||b - Tu||^2.$$

Therefore as in (2.1.1) we have

$$||Tu - b||^2 \geq ||Tu - Qb||^2 + ||b - Tu||^2$$

which gives

$$Tu - b = Qb - b \in \overline{R(T)}^{\perp} = N(T^*),$$

i.e. $T^*Tu = T^*b.$ #

<u>Definition</u>. A vector $u \in H_1$ which satisfies the equivalent conditions (i)-(iii) of (3.1.1) is called a <u>least squares solution</u> of the equation $Tx = b$.

Note that since we do not assume that $R(T)$ is closed there is no least squares solution of (1) if $Qb \notin R(T)$. Of course, for any vector b in the dense subspace

$$R(T) \oplus R(T)^{\perp} = \{y + z : y \in R(T), z \in R(T)^{\perp}\}$$

of H_2 equation (2) is solvable and we see from (3.1.1) (iii) that the set of all least squares solutions forms a closed convex set and hence has an element of smallest norm.

<u>Definition</u>. Let $T \in L(H_1, H_2)$. The mapping $T^\dagger$ with domain

$$D(T^\dagger) = R(T) \oplus R(T)^\perp$$

defined for $b \in D(T^\dagger)$ by $T^\dagger b = u$, where u is the least squares solution of minimal norm of the equation $Tx = b$, is called the <u>generalized inverse</u> of T.

If $R(T)$ is closed then clearly $D(T^\dagger) = H_2$ and the definition reduces to our previous notion of the generalized inverse. It can easily be shown as in the previous chapter that $T^\dagger$ is linear, $N(T^\dagger) = R(T)^\perp$, and $R(T^\dagger) = N(T)^\perp$, but it is no longer the case that $T^\dagger$ is necessarily bounded.

<u>Theorem 3.1.2</u>. If $T \in L(H_1, H_2)$, then $T^\dagger$ is bounded if and only if $R(T)$ is closed.

<u>Proof</u>. If $R(T)$ is closed, then $T^\dagger$ is bounded by (2.1.3). Suppose $R(T)$ is not closed, then

(3) $$TT^\dagger y = Qy$$

for each $y \in D(T^\dagger)$ and if $T^\dagger$ is bounded it has a unique continuous extension $\hat{T}^\dagger$ to $\overline{D(T^\dagger)} = H_2$, it then follows from (3) that

$$T\hat{T}^\dagger y = Qy, \quad y \in H_2.$$

But this surely cannot hold if $y \in \overline{R(T)}$ but $y \notin R(T)$. #

ARBITRARY RANGE CASE

We close this section with an important example which shows that the extension of the concept of the generalized inverse to bounded linear operators with nonclosed range is not merely generalization for its own sake. Consider a square integrable function

$$k(s,t) \in L^2([a,b] \times [a,b])$$

and define the integral operator

$$T : L^2[a,b] \to L^2[a,b]$$

by

(4) $$Tx(s) = \int_a^b k(s,t)x(t)dt, \quad s \in [a,b].$$

Equation (1) for this type of operator occurs in many applications. For an excellent survey of computational methods for generalized inverses of integral operators see the paper of Kammerer and Nashed [3].

The operator T defined by (4) is an example of a compact operator (see Taylor [1, p. 277]).

Definition. An operator $T \in L(H_1, H_2)$ is called <u>compact</u> if for each bounded set $U \subset H_1$, the closure of the set $T(U)$ is compact in the norm topology for H_2.

If $k(s,t)$ is a kernel of finite rank, that is

$$k(s,t) = \sum_{i=1}^{m} g_i(s) h_i(t)$$

for certain functions g_i and h_i, then the range of the integral operator (4) is finite dimensional and the solution of equation (1) reduces to solving a linear algebraic system (see e.g. Smithies [1]). The next theorem, which shows that the range of the integral operator (4) is closed only if it is finite dimensional, indicates the importance of defining the concept of the generalized inverse for operators with nonclosed range.

Theorem 3.1.3. If $T \in L(H_1, H_2)$ is compact, then $R(T)$ is closed if and only if $R(T)$ is finite dimensional.

Proof. (Kammerer and Nashed [3]).

Suppose that T is compact and $R(T)$ is closed. Then $T^\dagger$ is a bounded linear operator defined on all of H_2 and $TT^\dagger$ is compact (being the composition of a bounded linear operator with a compact operator). But $TT^\dagger = Q$ and $Q|R(T) = I|R(T)$ and hence the identity operator on $R(T)$ is compact. Therefore $R(T)$ is finite dimensional by exercise (2). The converse is trivial as every finite dimensional subspace is closed. #

ARBITRARY RANGE CASE   119

SECTION 2

A REPRESENTATION THEOREM

We will now give a general representation theorem similar to (2.3.1) for the generalized inverse of a bounded linear operator with nonclosed range. Of course in this case, since the generalized inverse is not bounded, we shall only obtain representations which converge in the strong operator topology. Also, since $R(T^*)$ need not be closed, $T^*T|\overline{R(T^*)}$ is not necessarily invertible and Theorem (2.1.4) is of no use. Nevertheless we can study the operator $\tilde{T} = T^*T|\overline{R(T^*)}$ which satisfies the formal identity

(1) $$\tilde{T}^{-1}T^*b = T^{\dagger}b$$

for any $b \in D(T^{\dagger})$ in the sense that

$$\tilde{T}T^{\dagger}b = T^*b,$$

indeed, this follows from condition (iii) in Theorem 3.1.1.

In analogy with our discussion in section 3 of the previous chapter, we set $H = \overline{R(T^*)}$ and define $\tilde{T} \in L(H,H)$ by $\tilde{T} = T^*T|H$. Then $\sigma(\tilde{T}) \subset [0,\infty)$ and motivated by (1) we will attempt to approximate $T^{\dagger}b$ by vectors of the form $U_\beta(\tilde{T})T^*b$, where $\{U_\beta(x)\}$ is a family of continuous functions on $[0,\infty)$ satisfying

$$\lim_\beta U_\beta(x) = x^{-1}, \quad \text{for } x \neq 0$$

and

$$|x\, U_\beta(x)| \leq M$$

for all $x \in [0,\infty)$ where M is a constant. We will have use for the following simple lemma.

<u>Lemma 3.2.1.</u> If $T \in L(H_1, H_2)$ and $x \in H_1$, then $T^\dagger T x = P_{N(T)^\perp} x$.

<u>Proof.</u> Suppose $x = x_1 + x_2 \in N(T)^\perp \oplus N(T)$. Then

$$T^\dagger T x = T^\dagger T x_1 = u$$

where u is the solution of minimum norm of the equation $Tu = QTx_1 = Tx_1$, i.e. $u = x_1$. Therefore $T^\dagger T x = x_1 = P_{N(T)^\perp} x$. #

The proof of the next theorem (Groetsch and Jacobs [1]) uses some basic facts from measure and integration theory a complete discussion of which may be found in Royden [1]. The idea of the proof is contained in Groetsch [4] where the special case of an integral equation of the first kind is considered.

ARBITRARY RANGE CASE

Theorem 3.2.2. Suppose $T \in L(H_1, H_2)$ and let $\tilde{T} = T^*T|H$ where $H = \overline{R(T^*)}$. If $\{U_\beta(x)\}$ is a net of continuous real-valued functions on $[0, ||T||^2]$ such that $\{x\, U_\beta(x)\}$ is uniformly bounded and $\lim_\beta U_\beta(x) = x^{-1}$ for $x \neq 0$, then for each $b \in D(T^\dagger)$,

$$T^\dagger b = \lim_\beta U_\beta(\tilde{T}) T^* b.$$

Proof. Let $\{E_\lambda\}$ be the resolution of the identity for the Hilbert space $H = N(T)^\perp$ generated by the self-adjoint operator $\tilde{T}$. Note that $\sigma(\tilde{T}) \subset [0, ||T||^2]$ and hence the functions $\{U_\beta(x)\}$ are continuous on $\sigma(\tilde{T})$. Now if $b \in D(T^\dagger)$, then for some $x \in H$ and $y \in R(T)^\perp$,

$$b = Tx + y$$

and hence

$$\begin{aligned} T^\dagger b &= T^\dagger T x + T^\dagger y \\ &= T^\dagger T x = P_{N(T)^\perp} x = x. \end{aligned}$$

On the other hand, since $x \in H$, we have

$$\begin{aligned} U_\beta(\tilde{T}) T^* b &= U_\beta(\tilde{T}) T^* T x \\ &= U_\beta(\tilde{T}) \tilde{T} x \\ &= \int_0^{||T||^2} \lambda U_\beta(\lambda) dE_\lambda x. \end{aligned}$$

Therefore

$$||U_\beta(\tilde{T})T^*b - T^\dagger b|| = ||U_\beta(\tilde{T})\tilde{T}x - x||$$
$$\leq \int_0^{||T||^2} |\lambda U_\beta(\lambda) - 1| \, d\mu$$

where μ is the measure on $(-\infty, \infty)$ which assigns to each interval $(a,b]$ the measure $||(E_b - E_a)x||$. Since $|\lambda U_\beta(\lambda) - 1|$ is uniformly bounded and

$$\lim_\beta |\lambda U_\beta(\lambda) - 1| = \begin{cases} 1, & \text{for } \lambda = 0 \\ 0, & \text{for } \lambda > 0, \end{cases}$$

we have by the bounded convergence theorem,

$$\lim_\beta \int_0^{||T||^2} |\lambda U_\beta(\lambda) - 1| \, d\mu = \mu\{0\} = 0$$

(recall that $E_\lambda = 0$ for $\lambda \leq 0$) which completes the proof. #

We shall leave to the reader the routine task of verifying that if $\{S_\beta(x)\}$ are the functions defined in any of the examples (1)-(6) of section 3 of the previous chapter, then they also satisfy the hypotheses of Theorem (3.2.2). It follows that each of the representations and computational procedures in examples (1)-(6) gives a method of computing $T^\dagger b$ for any $b \in D(T^\dagger)$.

ARBITRARY RANGE CASE

In particular we note that $T_t b \to T^\dagger b$ as $t \to \infty$ for each $b \in D(T^\dagger)$, where

$$T_t b = \int_0^t \exp(-T^* T u) T^* b \, du.$$

Under the additional assumption that $Qb \in R(TT^*)$ Showalter and Ben-Israel [1] have derived the error bound

$$||T_t b - T^\dagger b||^2 \le \frac{||T^\dagger b||^2 ||(TT^*)^\dagger b||^2}{||(TT^*)^\dagger b||^2 + 2||T^\dagger b||^2 t}.$$

Again, if $Qb \in R(TT^*)$ then

$$||T_N b - T^\dagger b||^2 \le \frac{||T^\dagger b||^2 ||(TT^*)^\dagger b||^2}{||(TT^*)^\dagger b||^2 + \alpha(2||T||^{-2} - \alpha) N ||T^\dagger b||^2}$$

where $0 < \alpha < 2||T||^{-2}$ and $\{T_n\}$ is defined by

$$T_0 = \alpha T^*, \quad T_{n+1} = (I_1 - \alpha T^* T) T_n + \alpha T^*.$$

Finally, if $Qb \in R(TT^*)$ then the hyperpower methods satisfy

$$||T_N^p b - T^\dagger b||^2 \le \frac{||T^\dagger b||^2 ||(TT^*)^\dagger b||^2}{||(TT^*)^\dagger b||^2 + (p^N - 1)\alpha(2||T||^{-2} - \alpha) ||T^\dagger b||^2},$$

where $\{T_N^p\}$ is defined in example 4 of section 3 of the previous chapter (see Showalter and Ben-Israel [1]).

SECTION 3

STEEPEST DESCENT

We now take up the steepest descent method for approximating $T^{\dagger}b$ where $T \in L(H_1, H_2)$ has arbitrary range and $b \in H_2$. Kammerer and Nashed [1] proved the convergence of the method and gave an error bound under the assumption that $Qb \in R(TT^*)$. More recently McCormick and Rodrigue [1] gave a unified treatment of the steepest descent method and several other gradient methods. The proof of the convergence of the steepest descent method given below follows that of McCormick and Rodrigue.

Recall that the steepest descent sequence is defined by

$$x_{n+1} = x_n - \alpha_n r_n, \quad n = 0, 1, 2, \cdots$$

where

$$r_n = T^* T x_n - T^* b$$

and

$$\alpha_n = ||r_n||^2 / ||T r_n||^2.$$

Below the error vector e_n is defined by

$$e_n = x_n - T^{\dagger} b - P_{N(T)} x_0.$$

Lemma 3.3.1. For $n = 0, 1, 2, \ldots$ the vectors r_n and e_n are related in the following way:

ARBITRARY RANGE CASE

(a) $T^*Te_n = r_n$,

(b) $\alpha_n ||r_n||^2 \leq (r_n, e_n)$,

and

(c) $||e_{n+1}||^2 \leq ||e_n||^2 - ||T||^{-2}||Te_n||^2$.

Proof. Part (a) follows since

$$r_n = T^*Tx_n - T^*b$$
$$= T^*T(x_n - T^\dagger b + P_{N(T)}x_0) = T^*Te_n.$$

Also

$$||r_n||^2 = (r_n, T^*Te_n) = (Tr_n, Te_n)$$
$$\leq ||Tr_n|| \, ||Te_n||,$$

but

$$||Te_n||^2 = (e_n, T^*Te_n) = (e_n, r_n),$$

therefore

$$||r_n||^2 \leq ||Tr_n||(e_n, r_n)^{1/2}.$$

Part (b) now follows since

$$\alpha_n ||r_n||^2 = ||r_n||^4 ||Tr_n||^{-2} \leq (e_n, r_n).$$

Finally,

$$\|e_{n+1}\|^2 = (e_n - \alpha_n r_n, e_n - \alpha_n r_n)$$
$$= \|e_n\|^2 - 2\alpha_n(r_n, e_n) + \alpha_n^2 \|r_n\|^2$$
$$\leq \|e_n\|^2 - 2\alpha_n(r_n, e_n) + \alpha_n(r_n, e_n)$$
$$= \|e_n\|^2 - \alpha_n \|Te_n\|^2,$$

and since $\alpha_n = \|r_n\|^2 \|Tr_n\|^{-2} \geq \|T\|^{-2}$, we have

$$\|e_{n+1}\|^2 \leq \|e_n\|^2 - \|T\|^{-2} \|Te_n\|^2. \quad \#$$

Theorem 3.3.2. Suppose $T \in L(H_1, H_2)$ and $b \in D(T^\dagger)$. Then for any $x_0 \in H_1$, the steepest descent sequence $\{x_n\}$ defined above converges monotonically to $T^\dagger b + P_{N(T)} x_0$.

Proof. (McCormick and Rodrigue [1]). Since $\{\|e_n\|^2\}$ is monotonically decreasing there is a $d \geq 0$ such that $\lim_n \|e_n\|^2 = d$. We wish to show that $d = 0$. By part (c) of the lemma

$$\|Te_n\|^2 \leq \|T\|^2 (\|e_n\|^2 - \|e_{n+1}\|^2),$$

therefore $\lim_n \|Te_n\|^2 = 0$. Choose a subsequence $\{x_{n_q}\}$ such that

(1) $\quad \|Te_{n_q}\| \leq \|Te_n\| \quad$ for $\quad n \leq n_q$.

ARBITRARY RANGE CASE

Then for any p and q,

$$(2) \quad ||e_{n_p} - e_{n_q}||^2 \leq 2\left|(e_{n_p} - e_{n_q}, e_{n_p})\right| + \left|||e_{n_p}||^2 - ||e_{n_q}||^2\right|.$$

But for $p \geq q$ we have by (3.3.1) (a) and (1)

$$|(e_{n_p} - e_{n_q}, e_{n_p})| = \left|\sum_{i=n_q}^{n_p-1} \alpha_i (r_i, e_{n_p})\right|$$

$$\leq \sum_{i=n_q}^{n_p-1} \alpha_i (Te_i, Te_{n_p})$$

$$\leq \sum_{i=n_q}^{n_p-1} \alpha_i ||Te_i|| \, ||Te_{n_q}||$$

$$\leq \sum_{i=n_q}^{\infty} \alpha_i ||Te_i||^2,$$

but in the proof of part (c) of the lemma we saw that

$$\alpha_n ||Te_n||^2 \leq ||e_n||^2 - ||e_{n+1}||^2.$$

Therefore

$$\sum_{i=0}^{\infty} \alpha_i ||Te_i||^2 \leq \sum_{i=0}^{\infty} (||e_i||^2 - ||e_{i+1}||^2) = ||e_0||^2 - d,$$

and hence

$$\lim_{q \to \infty} |(e_{n_p} - e_{n_q}, e_{n_p})| = 0.$$

By (2) we then have

$$\lim_{q \to \infty} ||e_{n_p} - e_{n_q}||^2 = 0.$$

Thus $\{e_{n_q}\}$ is a Cauchy sequence and therefore has a limit $v \in H_1$. We certainly have $||v|| = d$ and $v \in N(T)^\perp$ (exercise (4)). But $||Tv||^2 = \lim_n ||Te_n||^2 = 0$, therefore $v \in N(T)$. Hence $||v|| = d = 0$, completing the proof. #

The following error bound for the steepest descent sequence was given by McCormick and Rodrigue [1]. Suppose that $(I_1 - P_{N(T)})x_0 \in R(T^*)$ and $Qb \in R(TT^*)$, then

$$||e_n||^2 \le \frac{||T||^2 ||z_0 - z^*||^2 ||e_0||^2}{||T||^2 ||z_0 - z^*||^2 + n||e_0||^2}, \quad n = 0,1,2,\cdots$$

where $e_n = x_n - T^\dagger b - P_{N(T)}x_0$, z_0 is the unique vector in $\overline{R(T)}$ satisfying $T^* z_0 = (I_1 - P_{N(T)})x_0$, and z^* is the unique vector in $\overline{R(T)}$ satisfying $T^* z^* = T^\dagger b$ (such a z^* exists by virtue of the assumption that $Qb \in R(TT^*)$).

Before leaving the steepest descent method we shall present the following theorem (Groetsch [1]) which gives additional information on the nature of the convergence of the steepest descent sequence (see exercise (8) for another curious feature of the steepest descent sequence). Recall that a sequence $\{z_n\}$ in a Hilbert space H is said to converge weakly to $z \in H$ if

ARBITRARY RANGE CASE

$$\lim_n (z_n, y) = (z, y)$$

for each $y \in H$.

Theorem 3.3.3. Suppose $T \in L(H_1, H_2)$ and $b \in H_2$. Then each weak limit point of the steepest descent sequence $\{x_n\}$ is a least squares solution of the equation $Tx = b$.

Proof. If z is a weak limit point of $\{x_n\}$, then given $y \in H_1$, there is a subsequence $\{x_{n_j}\}$ with

(3) $$\lim_j (z - x_{n_j}, y) = 0$$

and

(4) $$\lim_j (x_{n_j} - z, y - T^*Ty) = 0.$$

However,

$$(x_{n_j} - z, y - T^*Ty) = (x_{n_j} - z, y) - (T^*Tx_{n_j} - T^*Tz, y)$$

$$= (x_{n_j} - z - r_{n_j} + T^*Tz - T^*b, y).$$

Substituting this into (4) and adding to (3) we obtain

$$\lim_j (T^*Tz - T^*b - r_{n_j}, y) = 0$$

or

$$(T^*Tz - T^*b, y) = \lim_j (r_{n_j}, y).$$

But by (2.4.1), r_{n_j} converges strongly to zero (recall that no assumption on the range of T was made in (2.4.1)), therefore

$$(T^*Tz - T^*b, y) = 0$$

for all $y \in H_1$, i.e. $T^*Tz = T^*b$. Hence z is a least squares solution of $Tx = b$. #

SECTION 4
THE CONJUGATE GRADIENT METHOD

We now present a theorem of Kammerer and Nashed [2] which is concerned with approximating least squares solutions of $Tx = b$ by use of the conjugate gradient method, where T is a bounded linear operator with arbitrary range. Recall that the conjugate gradient method is given by

$$p_0 = r_0 = T^*Tx_0 - T^*b,$$
$$x_1 = x_0 - \alpha_0 p_0,$$

where

$$\alpha_0 = ||r_0||^2 / ||Tr_0||^2, \text{ and for } i = 1, 2, \cdots$$

ARBITRARY RANGE CASE

(1) $$r_i = T^*Tx_i - T^*b$$
$$= r_{i-1} - \alpha_{i-1}T^*Tp_{i-1},$$

where

(2) $$\alpha_{i-1} = (r_{i-1}, p_{i-1})/||Tp_{i-1}||^2,$$

(3) $$p_i = r_i + \beta_{i-1}p_{i-1},$$

(4) $$\beta_{i-1} = -(r_i, T^*Tp_{i-1})/||Tp_{i-1}||^2,$$

and finally

(5) $$x_{i+1} = x_i - \alpha_i p_i.$$

In this section the operator T^*T will often be denoted by A.

<u>Lemma 3.4.1.</u> The entities defined above satisfy the following identities:

(a) $||Tp_i|| \leq ||Tr_i||$,

(b) $\beta_i = ||r_{i+1}||^2/||r_i||^2$,

(c) $(p_i, p_k) = ||r_i||^2 ||p_k||^2/||r_k||^2$,

(d) $p_i = ||r_i||^2 \sum_{j=0}^{i} r_j/||r_j||^2$,

(e) $||r_i|| \leq ||p_i||$.

Proof.

(a). For $i = 0$, this is clearly true. If $i \geq 1$, then by (3) we have

$$Tr_i = Tp_i - \beta_{i-1} Tp_{i-1}.$$

But since

$$(Tp_i, Tp_{i-1}) = (p_i, Ap_{i-1}) = 0$$

by (2.5.4), it follows by the Pythagorean property that

$$||Tr_i||^2 = ||Tp_i||^2 + \beta_{i-1}^2 ||Tp_{i-1}||^2$$
$$\geq ||Tp_i||^2.$$

(b). By (3) and (1) above and by (2.5.1) we have

$$||r_{i+1}||^2 = (r_{i+1}, p_{i+1})$$
$$= (r_{i+1}, r_i) - \alpha_i (r_{i+1}, Ap_i) + \beta_i (r_i, p_i)$$
$$- \alpha_i \beta_i (p_i, Ap_i).$$

By using (4) and (2.5.4) it follows that

$$||r_{i+1}||^2 = -\alpha_i (r_{i+1}, Ap_i) + \beta_i ||r_i||^2 + \alpha_i (r_{i+1}, Ap_i)$$
$$= \beta_i ||r_i||^2.$$

ARBITRARY RANGE CASE

Identity (c) was assigned as exercise (17) of Chapter II. Identity (d) is clearly true for i = 0. Also by (3) and (b), we have

$$p_{i+1} = r_{i+1} + \beta_i p_i$$
$$= r_{i+1} + ||r_{i+1}||^2 \sum_{j=0}^{i} r_j/||r_j||^2$$
$$= ||r_{i+1}||^2 \sum_{j=0}^{i+1} r_j/||r_j||^2$$

and (d) follows by induction.

The inequality (e) follows immediately from equation (21) in Section 5 of Chapter II and the Schwarz inequality.   #

We set $\bar{b} = Qb$, where Q is the projection of H_2 onto $\overline{R(T)}$ and define the functional g by

(6) $$g(x) = ||Tx - \bar{b}||^2.$$

Our aim is to approximate least squares solutions of Tx = b by minimizing the functional

$$f(x) = ||Tx - b||^2,$$

but note that since $f(x) = g(x) + ||b - \bar{b}||^2$, this is equivalent to minimizing g.

It will be shown below that under suitable conditions the conjugate gradient method converges to the least squares solution

$$u = T^\dagger b + (I_1 - P)x_0$$

of the equation $Tx = b$, where P is the projection of H_1 onto $\overline{R(T^*)}$. We define the error vector e_i by

$$e_i = x_i - u.$$

Note that since $T^*b = T^*\overline{b}$, and since u is a least squares solution, we have

(7) $\quad (r_i, e_i) = (T^*Tx_i - T^*\overline{b}, x_i - u)$

$\quad\quad\quad\quad\quad = (Tx_i - \overline{b}, Tx_i - Tu)$

$\quad\quad\quad\quad\quad = ||Tx_i - \overline{b}||^2 = g(x_i).$

Using (7) and the observation that $e_{i+1} = e_i - \alpha_i p_i$ and $Ae_i = r_i$, we obtain by (1), (2.5.1)(a) and (2.5.4)

(8) $\quad g(x_i) - g(x_{i+1}) = (r_i, e_i) - (r_{i+1}, e_i - \alpha_i p_i)$

$\quad\quad\quad\quad\quad\quad\quad = \alpha_i(Ap_i, e_i) + \alpha_i(r_{i+1}, p_i)$

$\quad\quad\quad\quad\quad\quad\quad = \alpha_i ||r_i||^2.$

We now present several lemmas proved by Kammerer and Nashed [2].

<u>Lemma 3.4.2.</u> (a) For $k = 0, 1, \ldots, i$,

ARBITRARY RANGE CASE

$$g(x_i) = (r_i, e_k) = (e_i, r_k).$$

(b) For $i = 0, 1, \ldots,$

$$(p_i, e_i)||r_i||^2 = g(x_i)||p_i||^2.$$

Proof.

(a). By successively using (5) we see that

(9) $$e_i = e_k - \sum_{j=k}^{i-1} \alpha_j p_j.$$

Therefore by (7) and (2.5.1)(a) we have

$$g(x_i) = (r_i, e_i) = (r_i, e_k).$$

Also

$$(r_i, e_k) = (Ae_i, e_k) = (e_i, Ae_k) = (e_i, r_k).$$

(b) To establish this identity we use (a), (3.4.1)(d) and the fact that the $\{r_i\}$ are orthogonal (see (2.5.4)). We therefore have

$$\begin{aligned}
(p_i, r_i)||r_i||^2 &= ||r_i||^4 \sum_{k=0}^{i} (r_k, e_i)/||r_k||^2 \\
&= g(x_i)||r_i||^4 \sum_{k=0}^{i} ||r_k||^{-2} \\
&= g(x_i)||p_i||^2. \quad \#
\end{aligned}$$

Lemma 3.4.3. For i = 0,1,...

$$||e_{i+1}||^2 \leq ||e_i||^2 - \alpha_i g(x_i).$$

Proof. By (9), (3.4.2)(b), (8), and (3.4.1)(e), we have

$$||e_{i+1}||^2 = ||e_i||^2 - 2\alpha_i(e_i,p_i) + \alpha_i^2||p_i||^2$$

$$= ||e_i||^2 - \alpha_i\{2g(x_i) - \alpha_i||r_i||^2\}||p_i||^2/||r_i||^2$$

$$= ||e_i||^2 - \alpha_i\{g(x_i) + g(x_{i+1})\}||p_i||^2/||r_i||^2$$

$$\leq ||e_i||^2 - \alpha_i g(x_i). \quad \#$$

Lemma 3.4.4. For all i and j, (p_i,e_i) and (e_i,e_j) are nonnegative.

Proof. Lemma (3.4.2)(b) shows that (p_i,e_i) is nonnegative; from this, (9), (3.4.1)(d), and (3.4.2)(a), it follows that (e_i,e_j) is nonnegative. #

Kammerer and Nashed establish the convergence of the conjugate gradient method under the assumption that

$$\bar{b} = Qb \in R(TT^*T).$$

Note that if this is so, then for some $\bar{z} \in R(T)$,

ARBITRARY RANGE CASE

(10) $$\bar{b} = TT^*\bar{z},$$

therefore

$$u = T^\dagger b = T^*\bar{z}$$

and by (10),

(11) $$\bar{z} = (TT^*)^\dagger b.$$

Also note that since $x_i \in R(T^*T)$ there exist vectors $z_i \in R(T)$ such that $x_i = T^*z_i$ and

(12) $$e_i = x_i - u = T^*(z_i - \bar{z}) \in R(T^*T).$$

Before proceeding to the next lemma it is necessary to introduce the concept of the adjoint of and unbounded linear operator. Suppose that $S : D(S) \to H_2$ is a linear operator defined on a dense subspace $D(S)$ of the Hilbert space H_1. It is easy to see that the set $D(S^*)$ defined by

$$D(S^*) = \{g \in H_2 : (Sf,g) = (f,h) \text{ for some } h \in H_1 \text{ and all } f \in D(S)\}$$

is a linear subspace of H_1. Also, by virtue of the fact that

$D(S)$ is dense, we see that the operator $S^* : D(S^*) \to H_1$ defined by $S^*g = h$ is well defined. This operator, called the <u>adjoint</u> of S, is clearly linear and satisfies

$$(Sf,g) = (f,S^*g)$$

for all $f \in D(S)$ and $g \in D(S^*)$. Note that in the case that S is a bounded operator this definition coincides with our previous notion of the adjoint.

If we define $W : \overline{R(T^*)} = N(T)^\perp \to R(T)$ by $W = T|\overline{R(T^*)}$, then we see that W is one to one and onto and therefore has an inverse W^{-1} defined on $R(T)$. Since $R(T)$ is not necessarily closed, W^{-1} is in general an unbounded linear operator. Similarly if $V = T^*|\overline{R(T)}$, then $V^{-1} : R(T^*) \to H_2$ exists as an unbounded linear operator. Note that if $x \in N(T)^\perp = \overline{R(T^*)}$ and $f \in R(T^*)$, say $f = T^*z$ where $z \in \overline{R(T)} = N(T^*)^\perp$, we have

$$(V^{-1}f,Tx) = (V^{-1}T^*z,Tx) = (z,Tx)$$

and

$$(f,W^{-1}Tx) = (f,x) = (T^*z,x) = (z,Tx).$$

Therefore we see that

$$(V^{-1}f,Tx) = (f,W^{-1}Tx)$$

for all $x \in \overline{R(T^*)}$, i.e. considering V^{-1} as an operator with

ARBITRARY RANGE CASE

domain in the Hilbert space $\overline{R(T^*)}$, we see that $R(T) \subset D(V^{-1*})$ and $V^{-1*}y = W^{-1}y$ for all $y \in R(T)$.

Lemma 3.4.5. If $Qb \in R(TT^*T)$, then for $i = 0,1,\cdots$

$$||e_{i+1}||^2 \leq (1 - B||e_i||^2)||e_i||^2,$$

where $B = ||z_0 - \bar{z}||^{-2}||T||^{-2}$.

Proof. Note that by (5) and the definition of z_i we have

$$z_i - z_{i+1} = \alpha_i V^{-1} p_i$$

where $V = T^* | \overline{R(T)}$. Therefore

$$\begin{aligned}
||z_{i+1} - \bar{z}|| &= ||z_i - \bar{z}||^2 - 2\alpha_i(V^{-1}p_i, z_i - \bar{z}) + \alpha_i^2 ||V^{-1}p_i||^2 \\
&= ||z_i - \bar{z}||^2 - \alpha_i \{2(V^{-1}p_i, z_i - \bar{z}) - (V^{-1}p_i, z_i - z_{i+1})\} \\
&= ||z_i - \bar{z}||^2 - \alpha_i(V^{-1}p_i, (z_i - \bar{z}) + (z_{i+1} - \bar{z})) \\
&= ||z_i - \bar{z}||^2 - \alpha_i(V^{-1}p_i, V^{-1}(e_i + e_{i+1}))
\end{aligned}$$

where in the last equality we have used (12). By use of the fact that $z_i - \bar{z} \in R(T)$ and the fact that $V^{-1*}y = W^{-1}y$ for $y \in R(T)$ where $W = T|\overline{R(T^*)}$, we find that

$$||z_{i+1} - \bar{z}||^2 = ||z_i - \bar{z}||^2 - \alpha_i(W^{-1}V^{-1}p_i, e_i + e_{i+1}).$$

Note that by (3.4.1)(a) we have

$$\alpha_i = ||r_i||^2/||Tp_i||^2 \geq ||r_i||^2/||Tr_i||^2 \geq ||T||^{-2}.$$

By exercise (10), $(W^{-1}V^{-1}p_i, e_i + e_{i+1}) \geq 0$, hence

$$||z_i - \bar{z}|| \leq ||z_i - \bar{z}|| \leq ||z_0 - \bar{z}||.$$

The Cauchy-Schwarz inequality then gives

$$\begin{aligned}||e_i||^4 &= |(x_i - u, T^*z_i - u)|^2 \\ &= |(T(x_i - u), z_i - \bar{z})|^2 \\ &\leq g(x_i)||z_0 - \bar{z}||^2.\end{aligned}$$

It now follows from (3.4.3) that

$$\begin{aligned}||e_{i+1}||^2 &\leq ||e_i||^2 - \alpha_i g(x_i) \\ &\leq ||e_i||^2 - \alpha_i ||e_i||^4/||z_0 - \bar{z}||^2 \\ &\leq (1 - B||e_i||^2)||e_i||^2.\end{aligned}$$

We now need a final lemma due to Kammerer and Nashed [2] (see also Showalter and Ben-Israel [1]).

Lemma 3.4.6. Let $\{C_i\}$ be a sequence of nonnegative numbers such that

ARBITRARY RANGE CASE

$$C_{i+1} \leq (1 - BC_i)C_i, \quad i = 0,1,\ldots$$

where $B > 0$ and $0 < BC_0 < 1$, then $C_{i+1} < C_i$ and

$$C_i \leq \frac{C_0}{1 + iBC_0}, \quad i = 0,1,\ldots.$$

Proof. Clearly we may assume that $C_k > 0$ for $k = 0,1,\cdots$. From the hypothesis we see that $C_{k+1} < C_k$ and

$$\frac{1}{C_{k+1}} - \frac{1}{C_k} = \frac{C_k - C_{k+1}}{C_k C_{k+1}} \geq \frac{BC_k^2}{C_k C_{k+1}} > B$$

for $k = 0,1,\cdots$. Therefore

$$\frac{1}{C_i} - \frac{1}{C_0} = \sum_{k=0}^{i-1} \left(\frac{1}{C_{k+1}} - \frac{1}{C_k}\right) > iB.$$

Solving this inequality for C_i we find

$$C_i \leq \frac{C_0}{1 + iBC_0}, \quad i = 0,1,2,\cdots. \quad \#$$

Theorem 3.4.7. Suppose $T \in L(H_1, H_2)$ where H_1 and H_2 are real Hilbert spaces. If $Qb \in R(TT^*T)$, then the conjugate gradient method with $x_0 \in R(T^*T)$ converges monotonically to $u = T^\dagger b$. Moreover, for a certain constant B,

$$\|x_i - u\|^2 \leq \frac{\|x_0 - u\|^2}{1 + iB\|x_0 - u\|^2}, \quad i = 1,2,\cdots.$$

Proof. (Kammerer and Nashed [2]). By (3.4.5),

$$||e_{i+1}||^2 \le (1 - B||e_i||^2)||e_i||^2,$$

where $B = ||z_0 - \bar{z}||^{-2}||T||^{-2}$. Also,

$$B||e_0||^2 = B||x_0 - u||^2 = B||T^*(z_0 - \bar{z})||^2$$
$$\le ||T^*||^2||T||^{-2} = 1.$$

Hence we may apply (3.4.6) with $C_i = ||e_i||^2$ to complete the proof. #

ARBITRARY RANGE CASE

EXERCISES

1. Show that $T \in L(H_1, H_2)$ is compact if and only if for each weakly convergent sequence $\{x_n\} \subset H_1$, the sequence $\{Tx_n\} \subset H_2$ is norm convergent. (Hint: Use the fact that each bounded sequence in a Hilbert space has a weakly convergent subsequence.)

2. Prove that an inner product space is finite dimensional if and only if the identity operator on the space is compact.

3. Supply the details which verify the inequality

$$||U_\beta(\tilde{T})\tilde{T}x - x|| \leq \int_0^{||T||^2} |\lambda U_\beta(\lambda) - 1| \, d\mu$$

   in the proof of (3.3.2).

4. Suppose $\{e_n\}$ is the sequence defined in section 3. Show that $e_n \in N(T)^\perp$ for $n = 0, 1, 2, \cdots$.

5. Suppose that $T \in L(H_1, H_2)$ and $b \in H_2$. Show that if for some $x_0 \in H_1$ the steepest descent sequence is bounded then the equation $Tx = b$ has a least squares solution.

   In the next three exercises (taken from Langlois [1]) we assume that $T \in L(H_1, H_2)$ and $b \in H_2$.

6. Suppose $z \notin N(T)$, $\varepsilon = ||z||^2 ||Tz||^{-2}$, and $w = z - \varepsilon T^*Tz$. Show that $w = 0$ if and only if z is an eigenvector of T^*T.

7. Show that the vector w defined in exercise (6) is not an eigenvector of T^*T.

8. Show that the steepest descent sequence to $Tx = b$ either converges at the first step or converges only in the limit as $n \to \infty$. (Consider the two cases where r_0 is an eigenvector of T^*T and where r_0 is not an eigenvector of T^*T and use exercises (6) and (7)).

9. Show that if $T \in L(H_1, H_2)$, then $T^\dagger$ is a <u>closed</u> operator, i.e. $\{b_n\} \subset D(T^\dagger)$, $b_n \to b$, and $T^\dagger b_n \to x$ imply that $b \in D(T^\dagger)$ and $T^\dagger b = x$.

10. Show that $(W^{-1}V^{-1}p_i, e_{i+1} + e_i) \geq 0$ for $i = 0, 1, 2, \ldots$, where V and W are defined as on page 138 and p_i and e_i are given in section 4. (Hint: Use (3.4.1)(d) and (3.4.4)).

BIBLIOGRAPHY

R. P. Agnew

[1] The Lototsky method of evaluation of series, Michigan Math. J. 4(1957), 105-128.

A. Albert

[1] Regression and the Moore-Penrose Pseudoinverse, Academic Press, New York, 1972.

J. Albrecht

[1] Bemerkungen zum Iterationsverfahren von Schulz zur Matrixinversion, Z. Angw. Math. Mech. 41(1961), 262-263.

M. Altman

[1] Approximation Methods in Functional Analysis, Lecture Notes Ma 107c, California Institute of Technology, Pasadena, 1959.

[2] An optimum cubically convergent iterative method of inverting a bounded linear operator in Hilbert space, Pacific J. Math. 10(1960), 1107-1114.

R. Ansorge

[1] Über ein Iterationsverfahren von G. Schulz zur Ermittlung der Reziproken einer Matrix, Z. Angew. Math. Mech. 39(1959), 164-165.

H. A. Antosiewicz and W. C. Rheinboldt

[1] Numerical analysis and functional analysis, in "A Survey of Numerical Analysis" (Ed. J. Todd), McGraw-Hill, New York, 1962.

E. Arghiriade

[1] Sur l'inverse généralisée d'un operatur linéaire dans les espaces de Hilbert, Atti. Acad. Naz. Lincei Rend. Cl. Sci. Fis. Mat. Natur., Ser. 8, 45(1968), 471-477.

E. Arghiriade and A. Dragomir

[1] Remarques sur quelques théorems relatives l'inverse généralisée d'un opératur linéaire dans les espace de Hilbert, Atti. Acad. Naz. Lincei Rend. Cl. Sci. Dis. Mat. Natur., Ser. 8, 46(1969), 333-338.

E. Arghiriade and E. Boros

[1] L'inverse généralisée d'un operatur linéaire dans un espace a produit interieur, Atti. Acad. Naz. Lincei Rend. Cl. Sci. Fis. Mat. Natur., Ser. 8, 46(1969), 646-649.

A. V. Balakrishnan

[1] An operator theoretic formulation of a class of control problems and a steepest descent method of solution, SIAM, J. Control 1(1963), 109-127.

BIBLIOGRAPHY

R. Bellman

    [1] A note on the summability of formal solutions of linear integral equations, Duke Math. J. 17(1950), 53-55.

A. Ben-Israel and A. Charnes

    [1] Contributions to the theory of generalized inverses. SIAM J. 11(1963), 667-697.

A. Ben-Israel and T. N. E. Greville

    [1] Generalized Inverses: Theory and Applications, Wiley, New York, 1974.

S. K. Berberian

    [1] Introduction to Hilbert Space, Oxford University Press, New York, 1961.

F. J. Beutler

    [1] The operator theory of the pseudo-inverse I. Bounded operators, J. Math. Anal. Appl. 10(1965), 451-470.

    [2] The operator theory of the pseudo-inverse II. Unbounded operators with arbitrary range, J. Math. Anal. Appl. 10(1965), 471-493.

H. Bialy

    [1] Iterative Behandlung linearer Functionalgleichungen, Ark. Rat. Mech. Anal. 4(1959), 166-176.

BIBLIOGRAPHY

A. Bjerhammer

[1] Rectangular reciprocal matrices with special reference to geodetic calculations, Bull. Géodésique 52(1951), 188-220.

R. Bouldin

[1] The pseudo-inverse of a product, SIAM J. Appl. Math. 24(1973), 489-495.

T. Boullion and P. Odell

[1] Generalized Inverse Matrices, Wiley, New York, 1971.

[2] Theory and Application of Generalized Inverses of Matrices, Symposium Proceedings, Texas Technological University, 1968.

S. L. Campbell and C. D. Meyer

[1] EP operators and generalized inverses, Canad. Math. Bull., to appear.

A. Cauchy

[1] Méthode générale pour la résolution des systémes d'équations simultanées, C. R. Acad. Sci. Paris 25(1847), 536-538.

S. R. Caradus

[1] Operator Theory of the Pseudo-Inverse, Queen's University Pure and Applied Mathematics Papers, No. 38, Kingston, Ontario, 1974.

BIBLIOGRAPHY

J. A. Clarkson

  [1] Uniformly convex spaces, Trans. Amer. Math. Soc. 40(1936), 396-414.

S. D. Conte and C. deBoor

  [1] Elementary Numerical Analysis, An Algorithmic Approach, Second Ed., McGraw-Hill, New York, 1972.

J. Daniel

  [1] The conjugate gradient method for linear and nonlinear operator equations, SIAM J. Numer. Anal. 4(1967), 10-26.

  [2] Approximate Minimization of Functionals, Prentice-Hall, Englewood Cliffs, N.J., 1971.

C. A. Desoer and B. H. Whalen

  [1] A note on pseudoinverses, SIAM J. 11(1963), 442-447.

W. Dück

  [1] Fehlerabschätzungen für das Iterationsverfahren von Schulz zur Bestimmung der Inverse einer Matrix, Z. Angew. Math. Mech. 40(1960), 192-194.

J. C. Dunn

  [1] Inversion of normal operators by polynomial interpolation, Proc. Amer. Math. Soc. 40(1973), 225-228.

BIBLIOGRAPHY

I. Erdelyi

[1] A generalized inverse for arbitrary operators between Hilbert spaces, Proc. Camb. Philos. Soc. 71(1972), 43-50.

I. Erdelyi and A. Ben-Israel

[1] Extremal solutions of linear equations and generalized inversion between Hilbert spaces, J. Math. Anal. Appl. 39(1972), 298-313.

D. K. Faddeev and V. N. Faddeeva

[1] Computational Methods of Linear Algebra, Freeman, San Francisco, 1963.

G. Forsythe and C. Moler

[1] Computer Solution of Linear Algebraic Systems, Prentice-Hall, Englewood Cliffs, N. J., 1967.

V. Fridman

[1] Method of successive approximations for Fredholm integral equations of the first kind, Uspehi Mat. Nauk. 11(1956), 233-234. (Russian)

[2] The convergence of the method of steepest descent, Uspehi Mat. Nauk 17(1962), 201-204. (Russian)

BIBLIOGRAPHY

C. W. Groetsch

[1] Steepest descent and least squares solvability. Canad. Math. Bull. 17(1974), 275-276.

[2] Representations of the generalized inverse, J. Math. Anal. Appl. 49(1975), 154-157.

[3] A product integral representation of the generalized inverse, Comment. Math. Univ. Carol. 16(1975), 13-20.

[4] On existence criteria and approximation procedures for integral equations of the first kind, Math. Comp. 29(1975), 1105-1108.

C. W. Groetsch and B. Jacobs

[1] A unified convergence theory for generalized inverses of bounded linear operators with arbitrary range, Notices A.M.S. 22(1975), A-681.

G. H. Hardy

[1] Divergent Series, Oxford University Press, London, 1949.

R. M. Hayes

[1] Iterative methods for solving linear problems in Hilbert space, N.B.S. Applied Mathematics Series, vol.39, U.S. Department of Commerce, Washington, D.C., 1954.

M. R. Hestenes

[1] The conjugate gradient method for solving linear

systems, Proc. Sixth Symp. Appl. Math., A.M.S., McGraw-Hill, New York, 1956.

M. R. Hestenes and E. Stiefel

[1] Method of conjugate gradients for solving linear systems, J. Res. Nat. Bur. Standards, Sec. B 49(1952), 409-432.

R. B. Holmes

[1] A Course in Optimization and Best Approximation, Lecture Notes in Mathematics, Vol. 257, Springer-Verlag, New York, 1972.

F. John

[1] Lectures on Advanced Numerical Analysis, Gordon and Breach, New York, 1967.

W. J. Kammerer and M. Z. Nashed

[1] Steepest descent for singular linear operators with nonclosed range, Applicable Anal. 1(1971), 143-159.

[2] On the convergence of the conjugate gradient method for singular linear operator equations, SIAM J. Numer. Anal. 9(1972), 165-181.

[3] Iterative methods for best approximate solutions of linear integral equations of the first and second kinds, J. Math. Anal. Appl. 40(1972), 547-573.

BIBLIOGRAPHY

L. V. Kantorovich and G. Akilov

[1] Functional Analysis in Normed Spaces, Pergamon, London, 1964.

K. Knopp

[1] Über das Eulersche Summierungsverfahren, Math. Z. 15(1922), 226-253.

J. J. Koliha

[1] The solution of linear operator equations in normed spaces by averaging iteration, SIAM J. Math. Anal. 5(1974), 178-186.

[2] Pseudo-inverses of operators, Bull. Amer. Math. Soc. 80(1974), 325-328.

W. E. Langlois

[1] Conditions for termination of the method of steepest descent after a finite number of iterations, I.B.M. J. Res. Develop. 10(1966), 98-99.

L. J. Lardy

[1] Some iterative methods for linear operator equations with applications to generalized inverses, Tech. Report TR 73-65, Depart. of Math., Univ. of Maryland, November, 1973.

[2] A series representation for the generalized inverse of a closed linear operator, Tech. Report TR 74-18, Depart. of Math., Univ. of Maryland, April, 1974.

A. T. Lonseth
[1] Approximate solutions of Fredholm-type integral equations, Bull. Amer. Math. Soc. 60(1954), 415-430.

L. Liusternik and V. Sobolev
[1] Elements of Functional Analysis, Ungar, New York, 1961.

E. R. Lorch
[1] Spectral Theory, Oxford University Press, New York, 1962.
[2] The spectral theorem, in "Studies in Modern Analysis", (R.C. Buck, Ed.), Mathematical Association of America, 1962.

W. R. Mann
[1] Mean value methods in iteration, Proc. Amer. Math. Soc. 4(1953), 506-510.

S. F. McCormick and G. H. Rodrigue
[1] A unified approach to gradient methods for linear operator equations, J. Math. Anal. Appl. 49(1975), 275-285.

BIBLIOGRAPHY

G. M. Muller

[1] Linear iteration and summability I, Notices Amer. Math. Soc. 15(1968), 329.

[2] Linear iteration and summability II, Notices Amer. Math. Soc. 18(1971), 643.

[3] Linear iteration and summability III, Notices Amer. Math. Soc. 18(1971), 812.

E. H. Moore

[1] Abstract, Bull. Amer. Math. Soc. 26(1920), 394-395.

[2] General Analysis, Part I, Amer. Philos. Soc., Philadelphia, 1935.

M. Z. Nashed

[1] Steepest descent for singular linear operator equations, SIAM J. Numer. Anal., 7(1970), 358-362.

[2] Generalized inverses, normal solvability and iteration for singular equations, in "Nonlinear Functional Analysis (L.B. Rall. Ed.), Academic Press, New York, 1971.

[3] Differentiability and related properties of nonlinear operators: some aspects of the role of differentials in nonlinear functional analysis, in "Nonlinear Functional Analysis" (L.B.Rall, Ed.), Academic Press, New York, 1971.

[4] Generalized Inverses and Applications, Academic Press, New York, 1976.

M. Z. Nashed and H. Salehi
[1] Measurability of generalized inverses of random linear operators, SIAM J. Appl. Math. 25(1973), 681-691.

M. Z. Nashed and G. F. Votruba
[1] A unified approach to generalized inverses of linear operators: I. Algebraic, topological and projectional properties, Bull. Amer. Math. Soc. 80(1974), 825-830.
[2] A unified approach to generalized inverses of linear operators: II. Extremal and proximal properties, Bull. Amer. Math. Soc. 80(1974), 831-835.

W. Niethammer and W. Schempp
[1] On the construction of iteration methods for linear equations in Banach spaces by summation methods, Aequationes Math. 5(1970), 273-284.

W. M. Patterson, 3rd
[1] Iterative Methods for the Solution of a Linear Operator Equation in Hilbert Space: A Survey, Lecture Notes in Mathematics, vol. 394, Springer-Verlag, New York, 1974.

BIBLIOGRAPHY

R. Penrose

[1] A generalized inverse for matrices, Proc. Cambridge Philos. Soc. 51(1955), 406-413.

[2] On best approximate solution of linear matrix equations, Proc. Cambridge Philos, Soc. 52(1956), 17-19.

W. V. Petryshyn

[1] On the inversion of matrices and linear operators, Proc. Amer. Math. Soc. 16(1965), 893-901.

[2] On generalized inverses and uniform convergence of $(I - \beta K)^n$ with applications to iterative methods, J. Math. Anal. Appl. 18(1967), 417-439.

R. Rado

[1] Note on generalized inverses of matrices, Proc. Cambridge Philos. Soc. 52(1956), 600-601.

C. R. Rao and S. K. Mitra

[1] Generalized Inverse of Matrices and Its Applications, Wiley, New York, 1971.

W. T. Reid

[1] Generalized inverses of differential and integral operators, in Boullion and Odell [2].

J. Reinermann

[1] Über Toeplitzsche Iterationsverfahren und einige Anwendungen in der konstruktiven Fixpunktheorie, Studia Math. 32(1969), 209-277.

F. Riesz and B. Sz.-Nagy

[1] Functional Analysis, Ungar, New York, 1956.

P. C. Rosenbloom

[1] The method of steepest descent, Proc. Sixth Symp. Appl. Math., A.M.S., McGraw-Hill, New York, 1956.

H. L. Royden

[1] Real Analysis, Macmillan, New York, 1963.

W. Schempp

[1] Iterative solution of linear operator equations in Hilbert space and optimal Euler methods, Arch. Math. 21(1970), 390-395.

G. Schulz

[1] Iterative berechnung der reziproken matrix, Z. Angew Math. Mech. 13(1933), 57-59.

R. D. Sheffield

[1] On Pseudo-Inverses of Linear Transformations in

BIBLIOGRAPHY

Banach Spaces, Oak Ridge National Laboratory Report, ORNL 2133, 1956.

[2] A general theory of linear systems, Amer. Math. Monthly 65(1958), 109-111.

M. Shönberg

[1] Sur la méthode d'iteration de Wiarda et Büchner pour la résolution de l'équation de Fredholm I, Acad. Roy. Belg. Bull. Cl. Sci. 37(1951), 1141-1156.

[2] Sur la méthode d'iteration de Wiarda et Büchner pour la résolution de l'équation de Fredholm II, ibid 38(1952), 154-167.

D. Showalter

[1] Representation and computation of the pseudoinverse, Proc. Amer. Math. Soc. 18(1967), 584-586.

D. Showalter and A. Ben-Israel

[1] Representation and computation of the generalized inverse of a bounded linear operator between two Hilbert spaces, Accad. Naz. dei Lincei 48(1970), 184-194.

E. Stiefel

[1] Some special methods of relaxation technique, N.B.S. Applied Mathematics Series, vol. 29, U.S. Department of Commerce, Washington, D.C., 1953.

F. Smithies
   [1]   Integral Equations, Cambridge University Press, Cambridge, 1970.

A. E. Taylor
   [1]   Introduction to Functional Analysis, Wiley, New York, 1958.

R. P. Tewarson
   [1]   On some representations of generalized inverses, SIAM Rev. 11(1969), 272-276.

A. N. Tihonov
   [1]   Regularization of incorrectly posed problems, Soviet Math. Doklady 4(1963), 1624-1627.

Yu. Tseng
   [1]   Sur les solutions des équations opératrices fonctionelles entre les espace unitaires. Solutions extrémales et solutions virtuelles, C.R. Acad. Sci. Paris 228(1948), 640-641.
   [2]   Generalized inverses of unbounded operators between two unitary spaces, Dokl. Akad. Nauk S.S.S.R. 67(1949), 431-434. (Russian)
   [3]   Properties and classification of generalized inverses of closed operators, ibid. 67(1949), 607-610. (Russian)

BIBLIOGRAPHY

    [4] Virtual solutions and generalized inverses, Uspehi Mat. Nauk 11(1956), 213-215. (Russian)

T. M. Whitney and R. K. Meany
    [1] Two algorithms related to the method of steepest descent, SIAM J. Numer. Anal. 4(1967), 109-118.

S. Zlobec
    [1] On computing the generalized inverse of a linear operator, Glasnik Mat. 2(22)(1967), 265-271.

INDEX OF SYMBOLS

$\lVert \cdot \rVert$	2
$(\cdot, \cdot)$	2
$\perp$	8
$\oplus$	9
$L(E_1; E_2)$	10
$R(T)$	11
$N(T)$	11
T^*	13
$T \vert S$	14
P_S	15
$\rho(T)$	17
$\sigma(T)$	18
E_λ	21
$f'(x_0)$	25
$\nabla f(x_0)$	29
$Df(x_0, h)$	30
$T^\dagger$	41
(V)	41
(M)	47
(P)	48
$(M - P)$	51
$(D - W)$	51
$S_\beta(A)$	56
E_α	63
$S_n^p(x)$	69
T_n^p	72
Δ	76
$\langle \cdot, \cdot \rangle$	96
$D(T^\dagger)$	116

SUBJECT INDEX

A	adjoint	13
B	Banach's Theorem	14
	Borel's Theorem	60
C	Cauchy sequence	4
	closed operator	144
	closed subspace	6
	compact operator	117
	complete space	4
	conjugate directions	91
	conjugate gradient method	
	operators with closed range	90
	operators with arbitrary range	130
	convex set	6
D	directional derivative	30
	dual space	5
E	eigenvalue	17
	eigenvector	17
	Euler method	66
	Euler-Knopp transform	62
F	Fréchet derivative	25
G	generalized inverse	
	variational definition	41
	Moore's definition	47
	Penrose's definition	48

	Desoer and Whalen's definition	51
	equivalence of definitions	52
	operators with arbitrary range	115
	gradient	29
H	Hilbert space	4
	hyperpower methods	71
I	infinite products	78
	inner product	2
	integral representation	59
	interpolation	76
K	kernel of finite rank	118
L	least squares solution	40,115
	Le Roy transform	82
	Lindelof transform	82
	linear functional	5
	linear operator	10
M	Mittag-Leffler transform	82
N	nonnegative operator	18
	normed linear space	1
	norm of a linear operator	10
	nullspace	11
O	orthogonal complement	8
P	parallelogram law	4
	positive operator	18
	projection	15
	Pythagorean property	8

SUBJECT INDEX

R	range	11
	resolution of the identity	21
	resolvent set	17
	Riesz Representation Theorem	6
S	Schulz's method	69
	Schwarz's inequality	3
	self-adjoint operator	13
	separable space	96
	Showalter's iterative method	62
	spectral mapping theorem	23
	spectral radius	18
	spectral theorem	22
	spectrum	18
	steepest descent	
	closed range case	82
	arbitrary range case	124
	convergence in finitely many steps	144
	strictly convex space	33
	strong convergence	11
	successive approximations	67
	summability	55
T	Tihonov regularization	74
U	uniform convergence	11
	uniformly convex space	33
W	weak convergence	128

QA
334.2
G76